D1254934

400 Full-Size Mini-Clock
P A T T E R N S

Rick & Karen Longabaugh

Acknowledgments & Credits

Rick & Karen Longabaugh
Lorna Smith Graphic Design
Owen & Owen Photography

ISBN-0-9633112-6-3
Copyright 1994 THE BERRY BASKET All Right Reserved
Printed in the USA

Contents

Introduction

Welcome to the wonderful world of clock making. This unique book features over 400 full-size pattern combinations for the popular 1 3/8" mini-clock inserts. Precise computer graphics and easy-to- follow instructions for each pattern will enable you to complete your project with professional results.

Included are projects for every skill level and pattern themes to please all:

- Country
- Victorian
- Religious
- Wildlife
- Sports
- Executive
- Children
- Southwest
- Miscellaneous
- Plus many more

These fantastic projects are great for gifts as well as making extra income. They're functional as well as decorative, compact in size and are a great way to turn your scrap wood into BIG PROFITS!

Instructions

TRANSFERRING THE PATTERN TO THE WORK PIECE

To use this pattern book most effectively, we suggest making photo copies of the patterns you wish to cut out. An advantage to the copier is that you can enlarge or reduce the pattern to fit the size of wood you choose to use. Use a spray adhesive to adhere the pattern to the wood. Spray adhesives can be purchased at most arts & crafts, photography, and department stores. Pay special attention to purchase one that states "temporary bond" or "repositionable". Lightly spray the back of the pattern, not the wood, then position the pattern onto the work piece.

SELECTING THE MATERIALS

You can select your material from a variety of plywoods or hardwoods. Plywoods such as Baltic Birch, lumber core, or furniture grade would work best. For hardwoods, any of the domestic or imported varieties will work well - ash, maple, walnut, oak birch mahogany, cherry, hickory, alder, pine, koa, etc.

The patterns in this book have been designed so that a variety of stock thicknesses can be used from 1/8" to 3/4". For the majority of patterns, however, 1/4" or 3/8" material would be ideal.

Once you have selected your material, cut to size the blanks for each piece of the clock you have chosen to make. Sand the surfaces beginning with #80 grit sandpaper, then #120, and finish with #220 grit.

VEINING

Veining is a simple technique that brings a "lifelike" appearance to many projects. For instance, the folds of clothing or the veins of a leaf take on a more realistic appearance when this technique is incorporated.

To vein, simply saw all black lines as indicated on the pattern. Some areas you will be able to vein by simply sawing inward from an outside edge, while to vein other areas you will need to drill a tiny blade entry hole, and then proceed to saw.

SILHOUETTE BACKING

Many of the projects cut from thin material can easily have a contrasting background added, which beautifully enhances the appearance of the design. This background offers versatility - it can be made from hardwood or plywood and it can be stained, painted, or even left natural.

To utilize this technique, first saw the fretwork portion of the project, leaving the outer shape uncut until you are ready to glue on the backing. Use a painting or cosmetic sponge to apply a thin layer of glue to the backside of the fretted work piece. Center this onto the backing blank and clamp until dry. Then simply cut along the outside line where indicated and use a 1/4"R roundover bit if desired. Sand where necessary.

CUTTING THE DESIGN

To begin, adhere the paper pattern to the work piece with a repositionable spray adhesive. If you are going to use a drill press with a 1 3/8" forstner bit to bore the hole for the clock insert, do so at this time. Be sure to clamp your work piece to the drill press table for safety. Bore 1/4" deep. If you are going to use your scroll saw to cut the opening for the clock insert, do so at this time. Follow along the solid line to cut the 1 3/8" opening.

Next proceed to cut the outer shape of the clock following the solid outside line.

PLEASE NOTE: If you have chosen to add a contrasting baking to your project please refer to the procedures outlined in the paragraph on silhouette backing at this time.

Now, drill starter holes where needed in the non-shaded areas and proceed to cut the design portion of the project. This is the time to use the veining technique to give the design its "lifelike" appearance. However, for an easier project, simply omit the veining.

CHOOSING THE BASE

Several styles and sizes of bases have been provided on pages 7-14. This allows you to choose the style and size of base to fit the particular project you are making.

Again, adhere the paper pattern to the work piece. Proceed by turning the work piece onto its side with the decorative pattern face up. Then cut along the solid outside line.

FINISHING AND ASSEMBLING

When all the pieces to a project have been cut, remove the paper patterns. Rout along the edges, if desired, with a small roundover bit. Then, sand any rough edges.

Assemble the clock by attaching it to the base, along with following any specific instructions that may be given on certain projects. These specific instructions can be found on the pattern page itself.

Where no specific instructions have been given for attaching the clock to the base, use one or more of the following after centering the clock on the base: glue, #18 x 5/8" finishing nails, an/or #6 x 3/4" (or smaller) flathead wood screws.

If any glue has been used, be sure to allow it to dry before finishing with a penetrating oil such as the Watco Danish Oils, Min-Wax Antique Oil Finish, Tung Oils, etc. An easy method is to pour the oil into a shallow pan and soak the project for approximately 30 seconds. Then, follow the oil manufacturer's directions on the container for finishing. When dry, insert the clock.

Helpful Tips

INSERTING THE CLOCK

If you used your scroll saw to cut the hole for the clock, and you find that upon inserting the clock that the opening is slightly too big, place a thin rubber band or other material around the insert to take up the slack.

VIDEO

If you would like to see a video produced on making these unique and fascinating clock projects, please use the form on page 127 to let us know.

SOURCES

To purchase the clock inserts call The Berry Basket at 1-800-206-9009 or visit us on the internet at WWW.BERRYBASKET.COM.

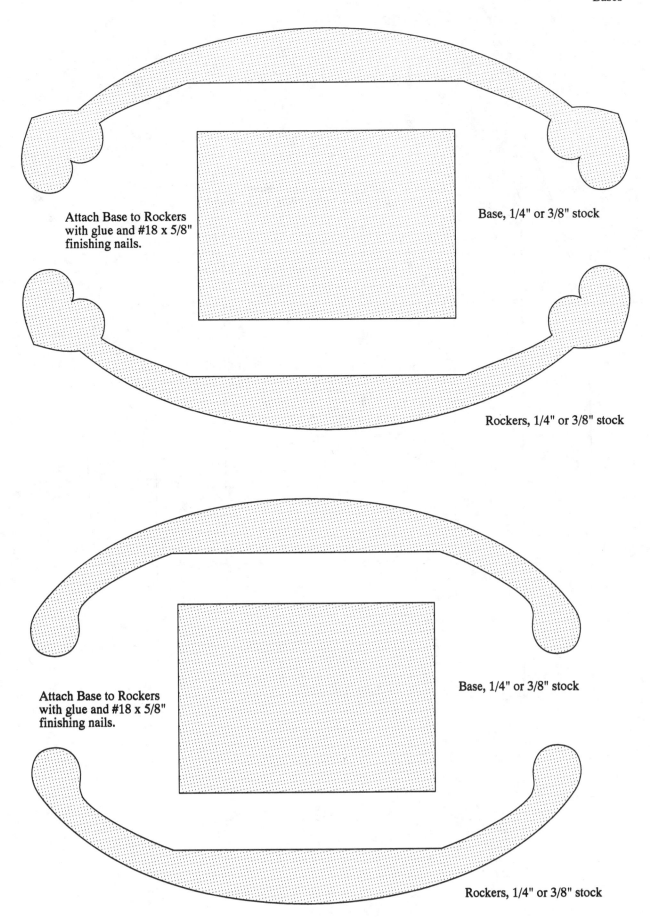

Attach Base to Rockers
with glue and #18 x 5/8"
finishing nails.

Base, 1/4" or 3/8" stock

Rockers, 1/4" or 3/8" stock

Attach Base to Rockers
with glue and #18 x 5/8"
finishing nails.

Base, 1/4" or 3/8" stock

Rockers, 1/4" or 3/8" stock

Attach Base to Rockers with glue
and #18 x 5/8" finishing nails.

Base, 1/4" or 3/8"
stock

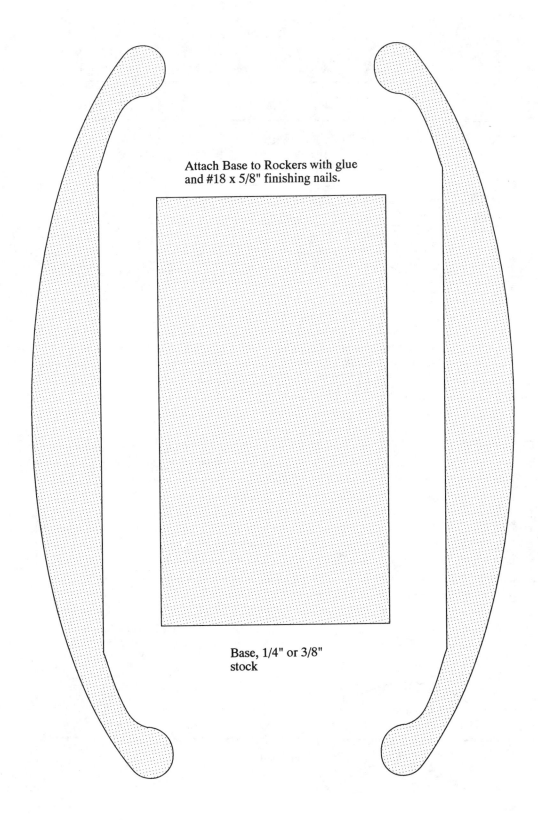

Attach Base to Rockers with glue
and #18 x 5/8" finishing nails.

Base, 1/4" or 3/8"
stock

Bases

Stock, 3/4"

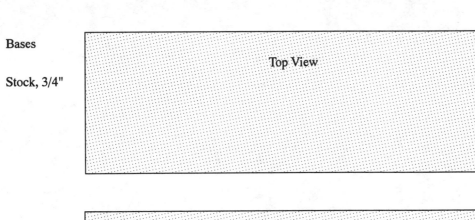

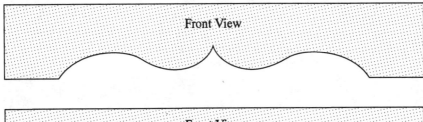

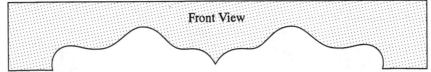

Cut the work piece to size according to the "Top View" dimensions.
Adhere the "Front View" pattern to the front of the work piece.
Proceed by turning the work piece onto its side with the decorative
pattern face up. Then cut along the solid outside line.

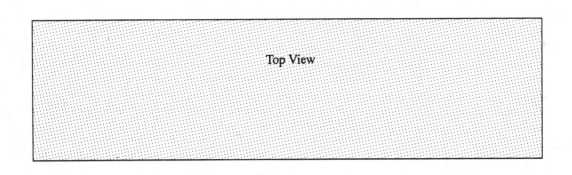

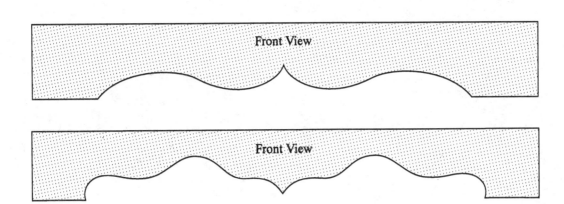

After centering clock on base, attach with glue
and finishing nails or flathead wood screws.

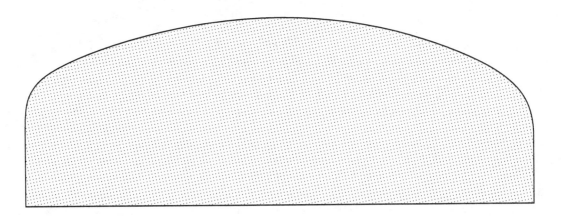

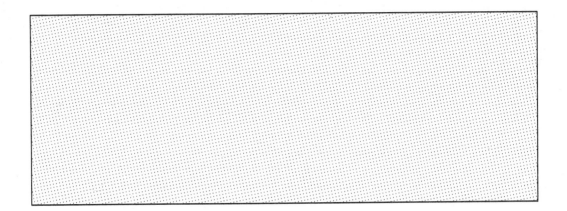

Bases

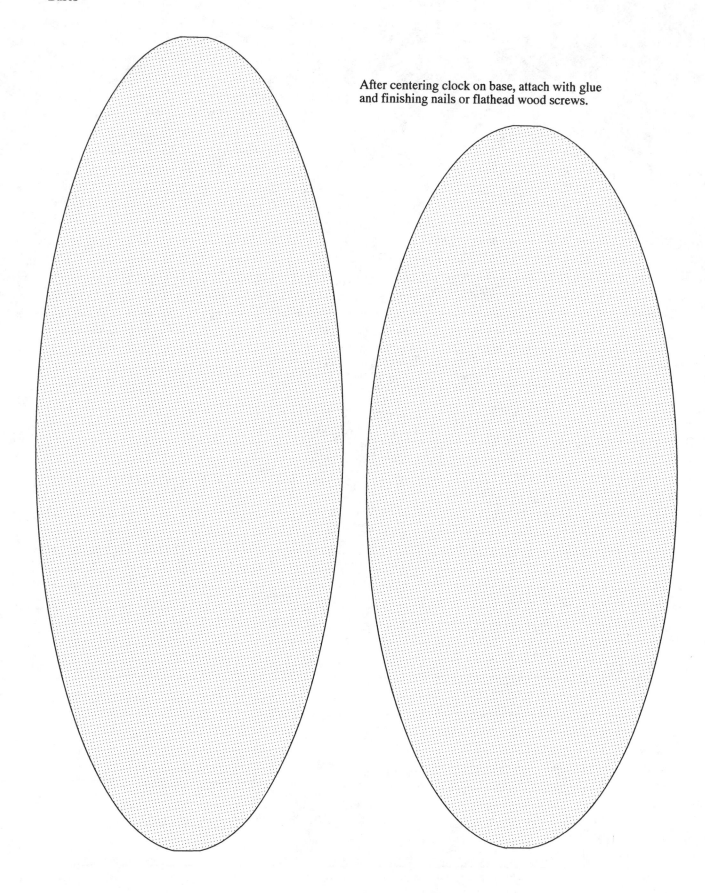

After centering clock on base, attach with glue
and finishing nails or flathead wood screws.

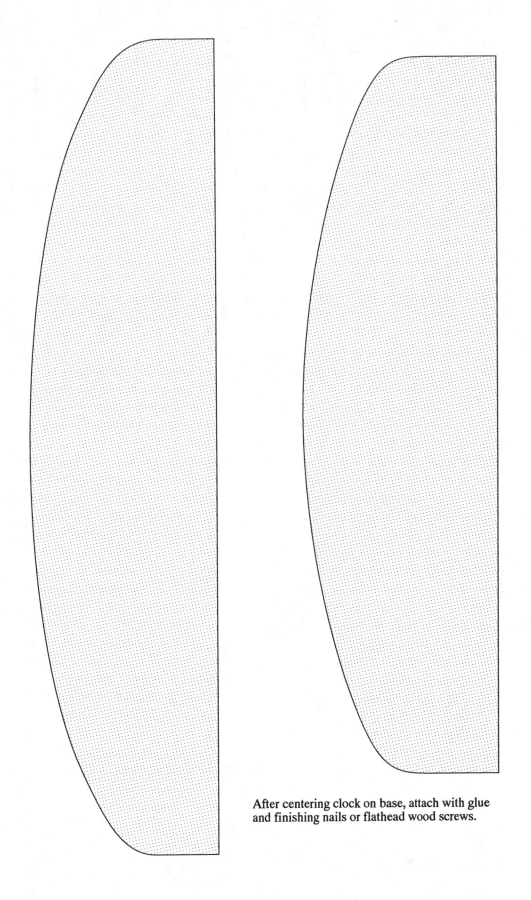

After centering clock on base, attach with glue and finishing nails or flathead wood screws.

After centering clock on base, attach with glue and finishing nails or flathead wood screws.

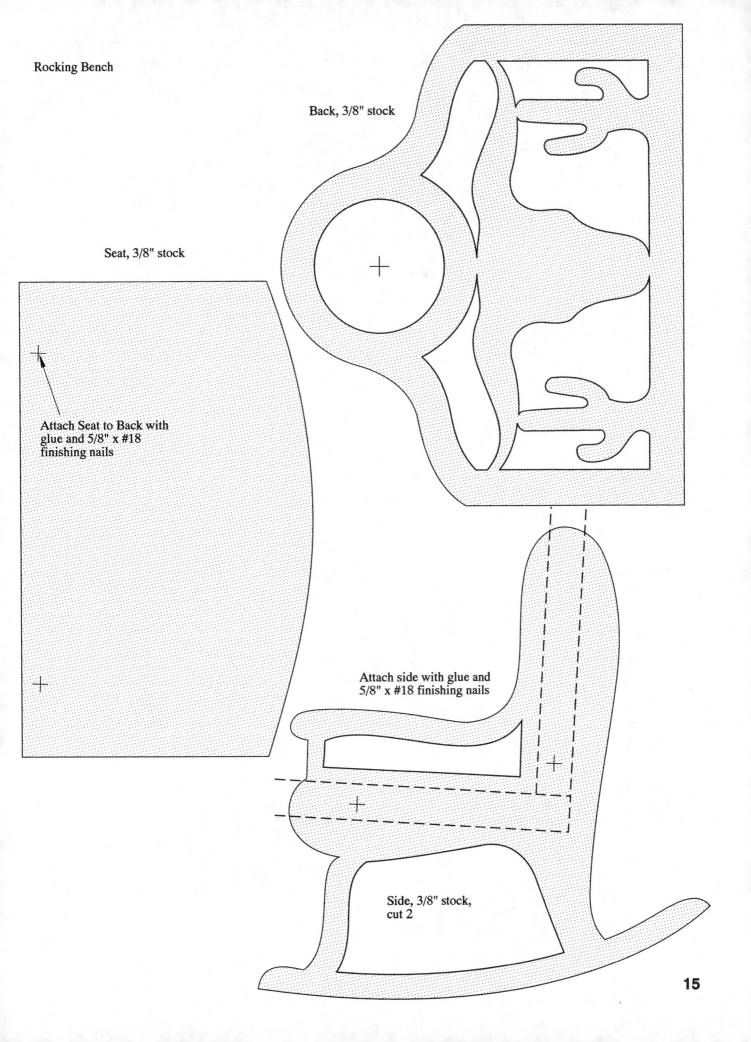

Rocking Bench

Back, 3/8" stock

Seat, 3/8" stock

Attach Seat to Back with glue and 5/8" x #18 finishing nails

Attach side with glue and 5/8" x #18 finishing nails

Side, 3/8" stock, cut 2

15

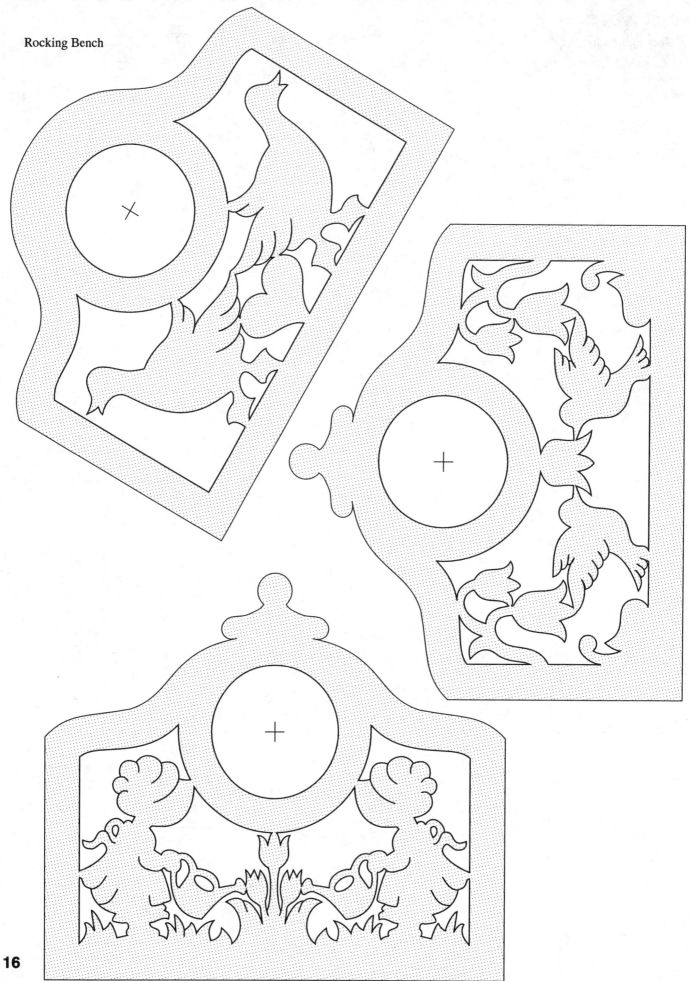

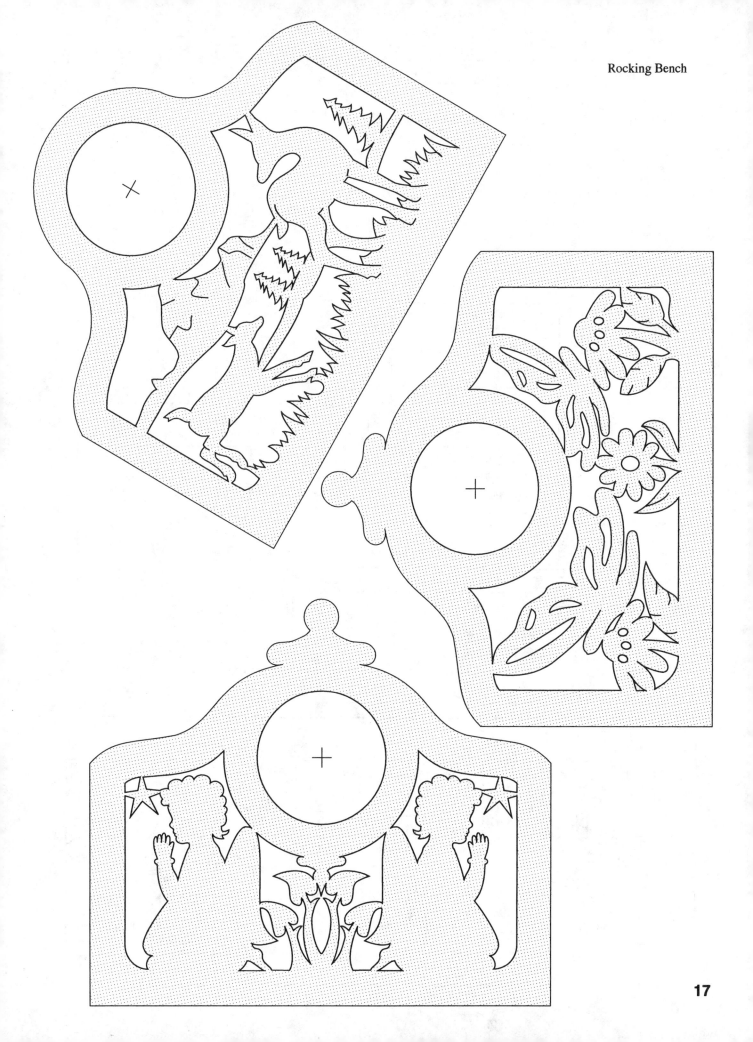

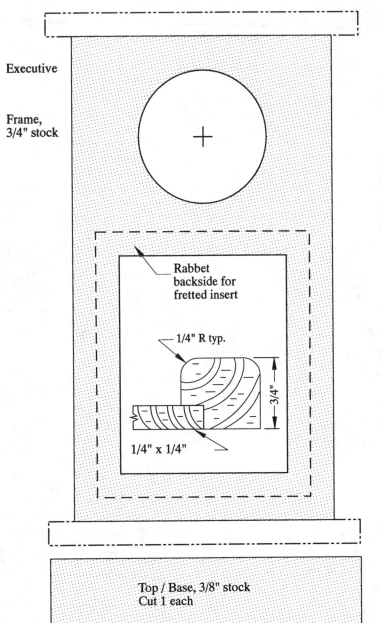

Executive

Frame,
3/4" stock

Rabbet
backside for
fretted insert

1/4" R typ.

3/4"

1/4" x 1/4"

Top / Base, 3/8" stock
Cut 1 each

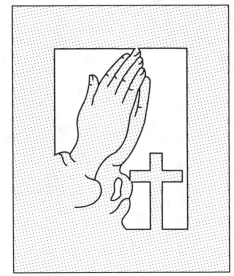

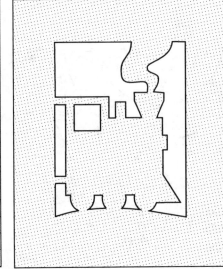

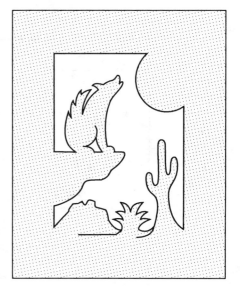

Executive

Frame and Base, 3/8" stock

Attach frame here

Attach frame here

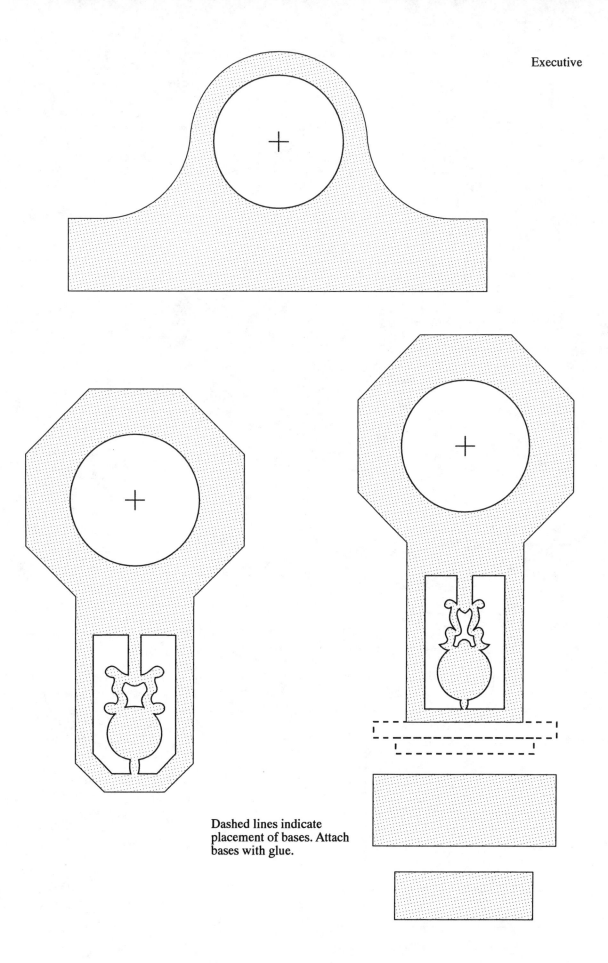

Dashed lines indicate placement of bases. Attach bases with glue.

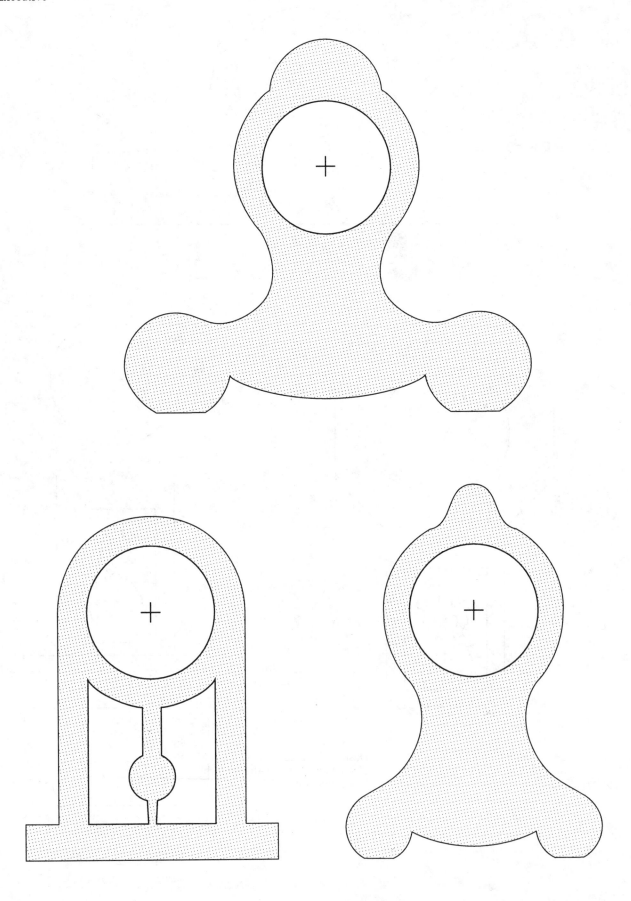

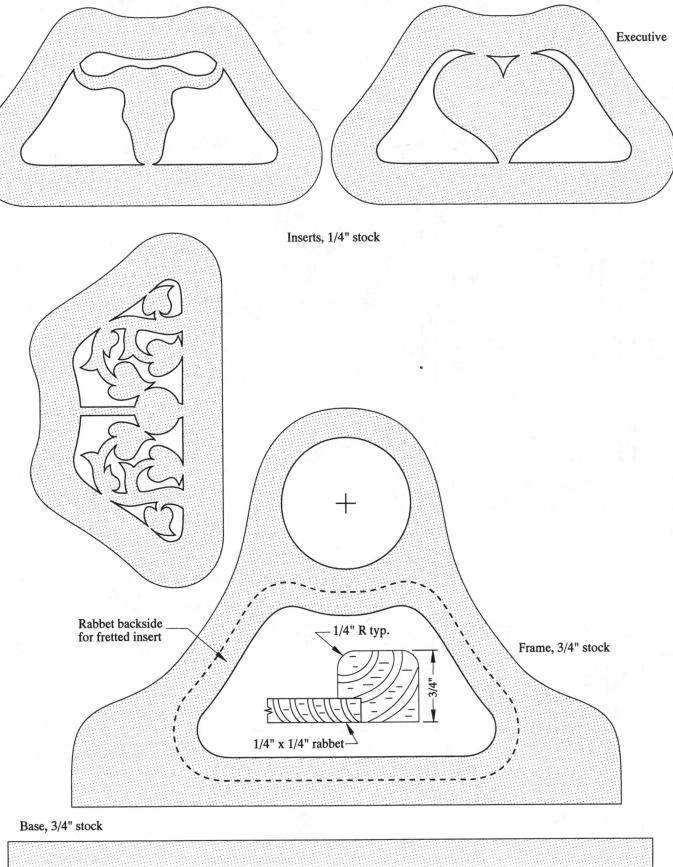

Executive

Inserts, 1/4" stock

Rabbet backside
for fretted insert

1/4" R typ.

3/4"

1/4" x 1/4" rabbet

Frame, 3/4" stock

Base, 3/4" stock

23

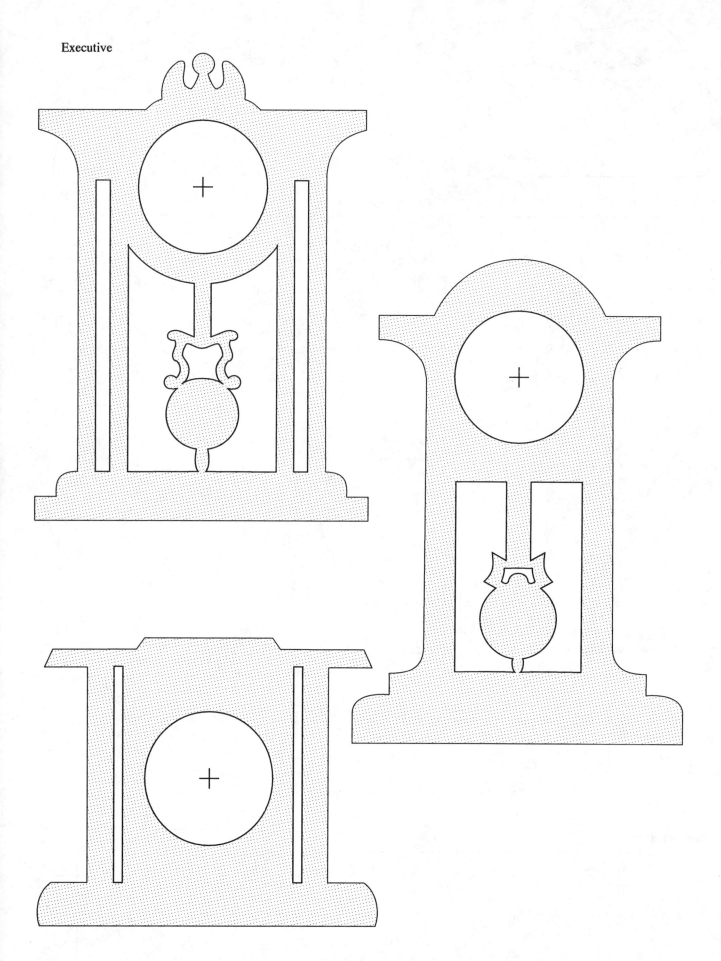

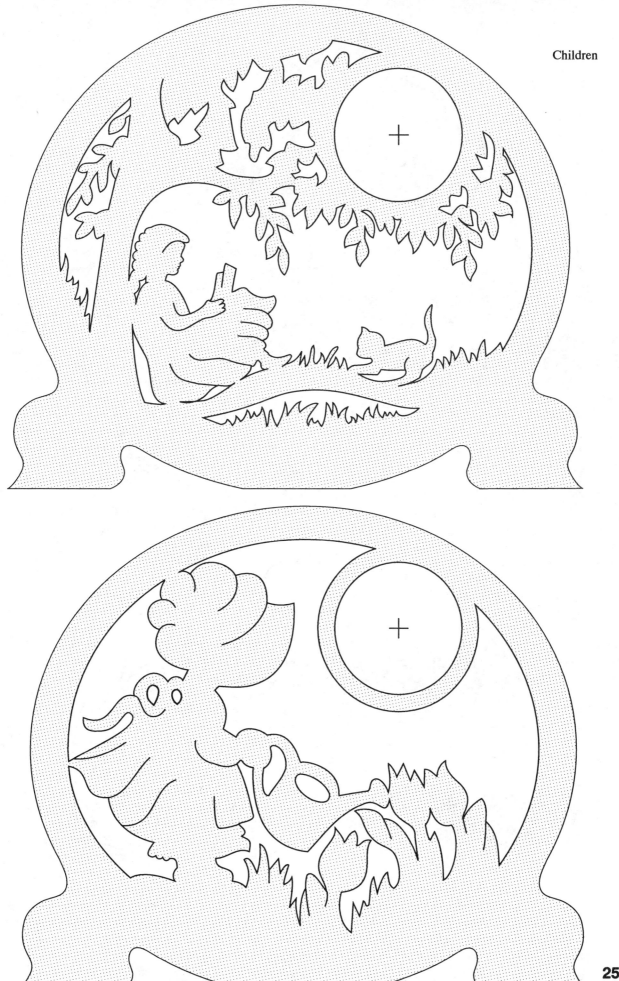

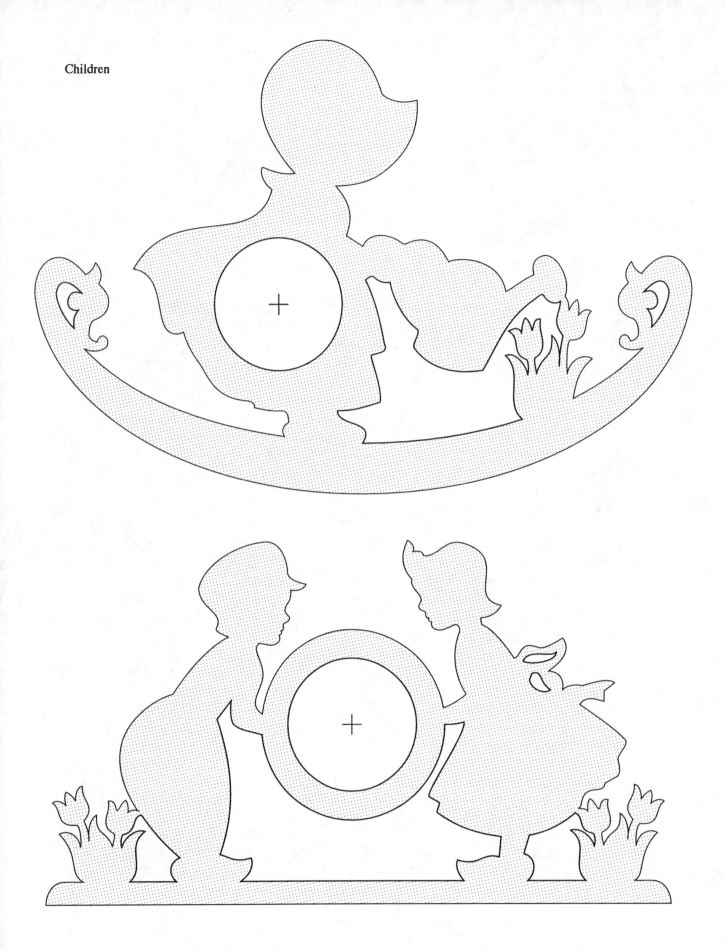

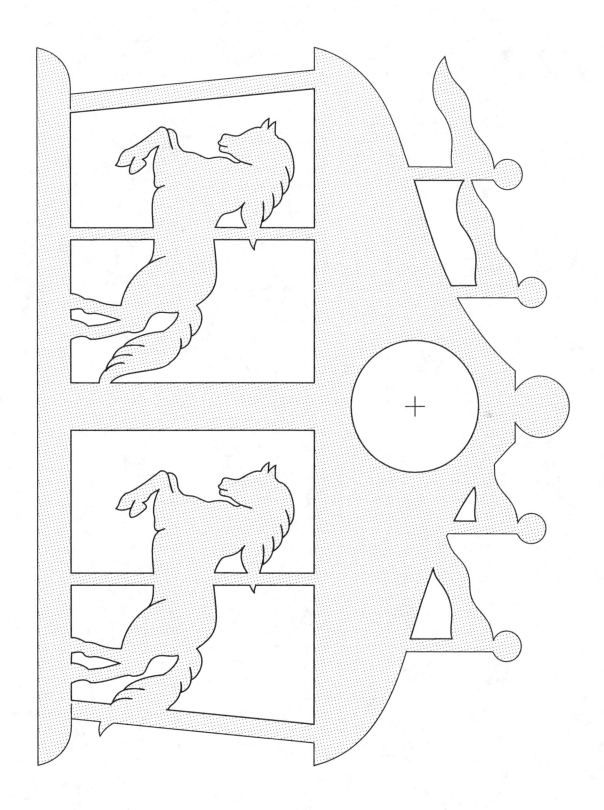

Children

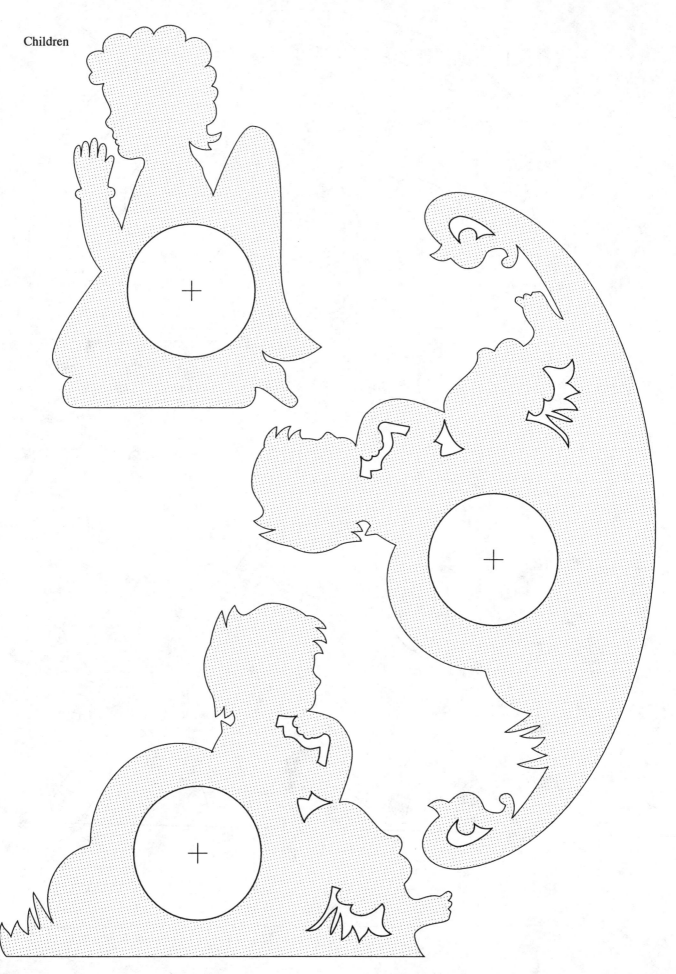

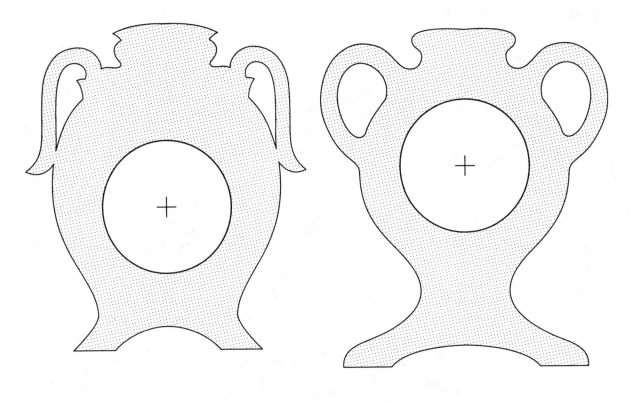

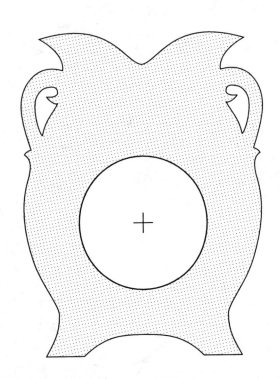

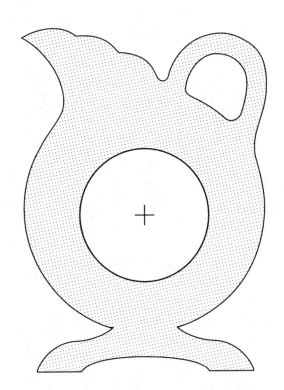

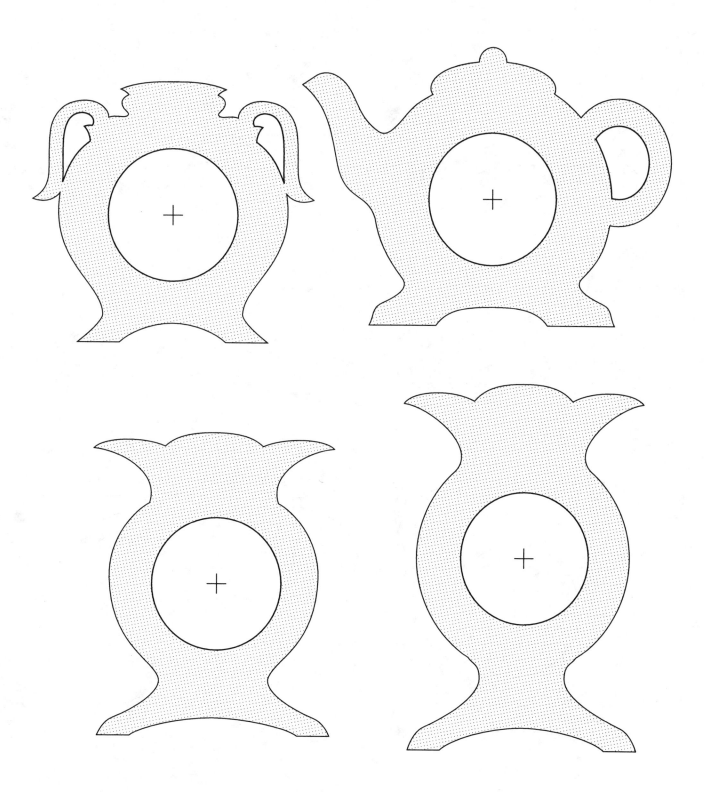

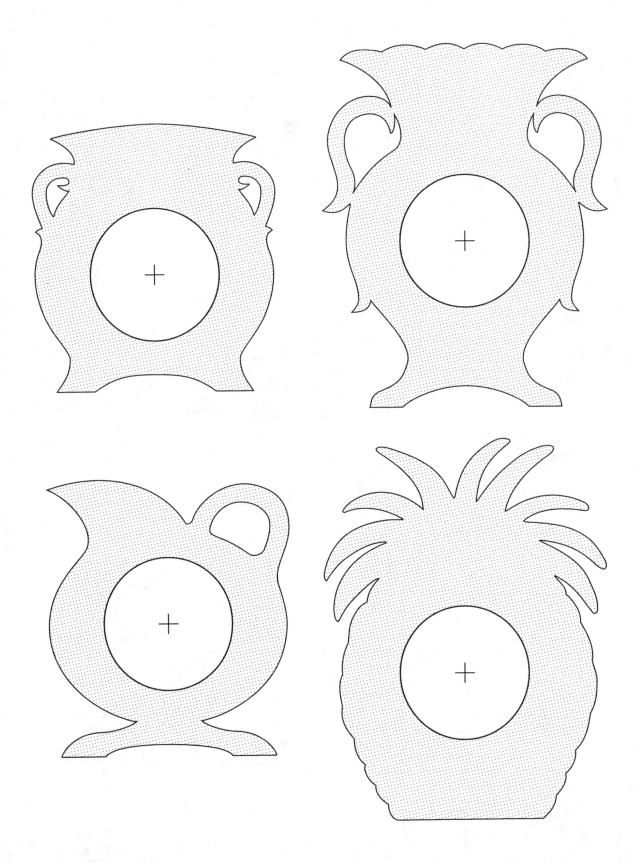

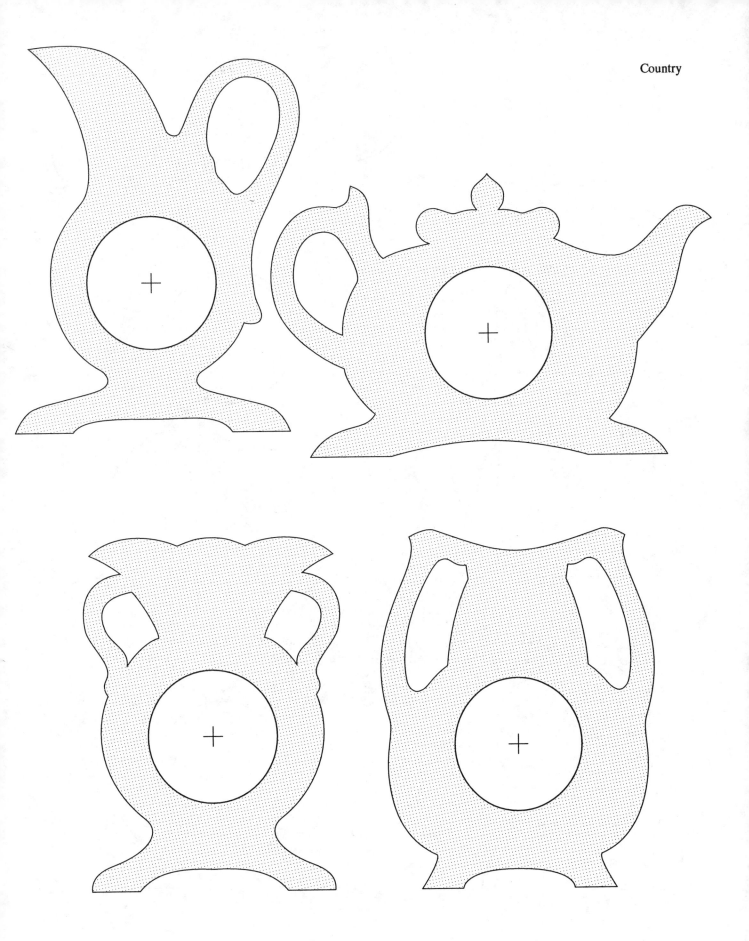

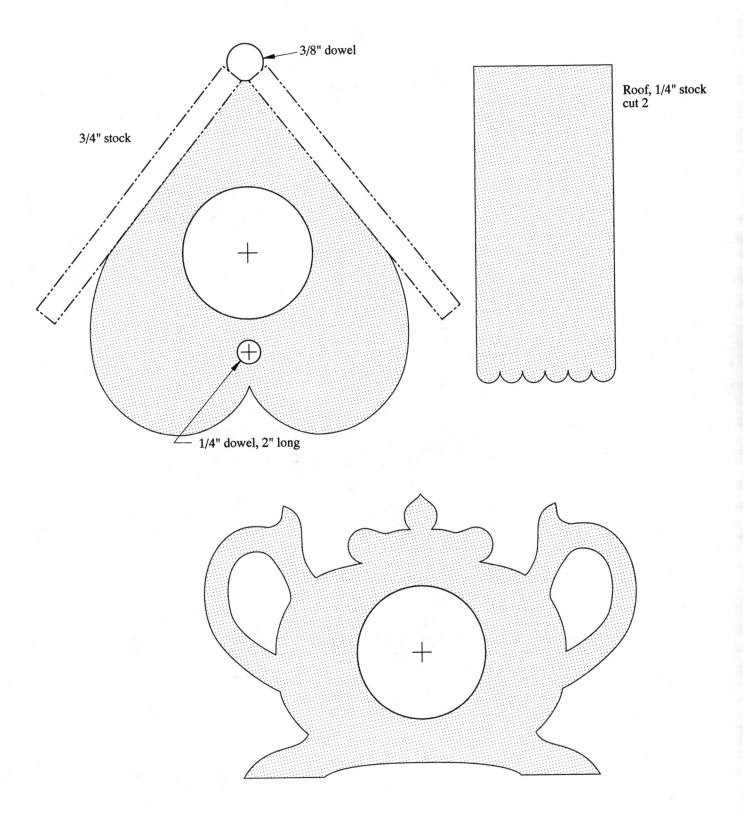

3/8" dowel

3/4" stock

Roof, 1/4" stock
cut 2

1/4" dowel, 2" long

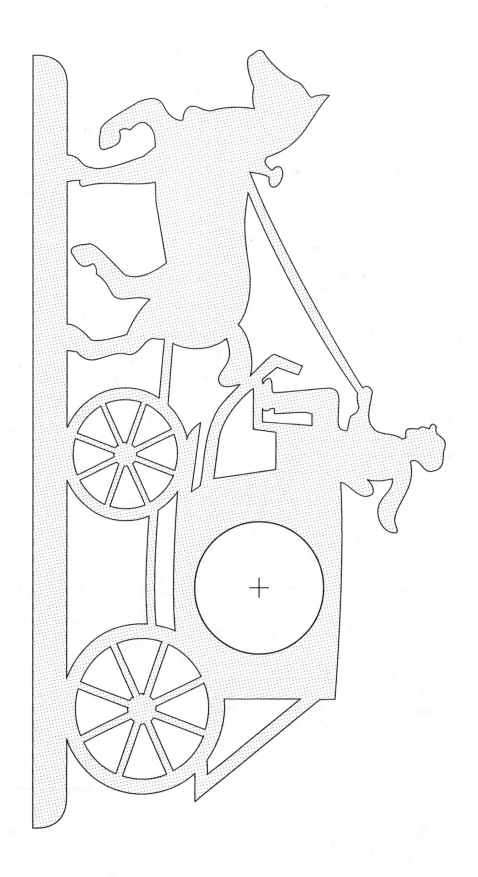

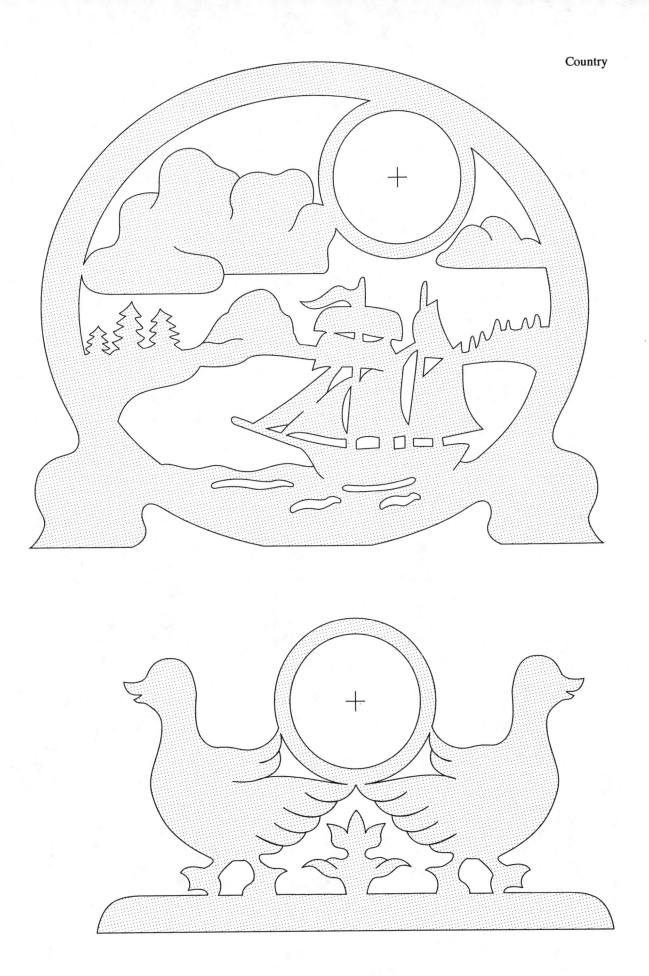

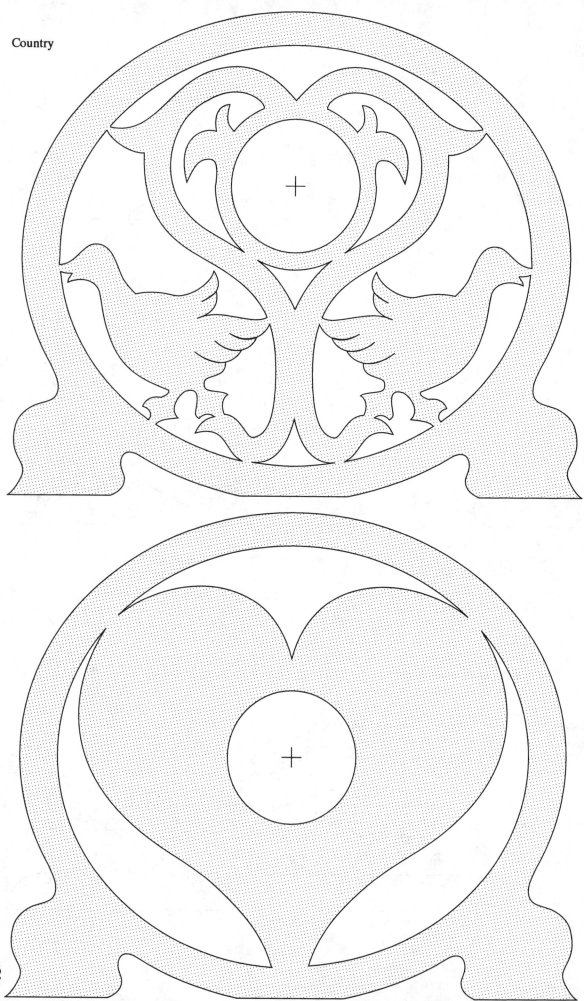

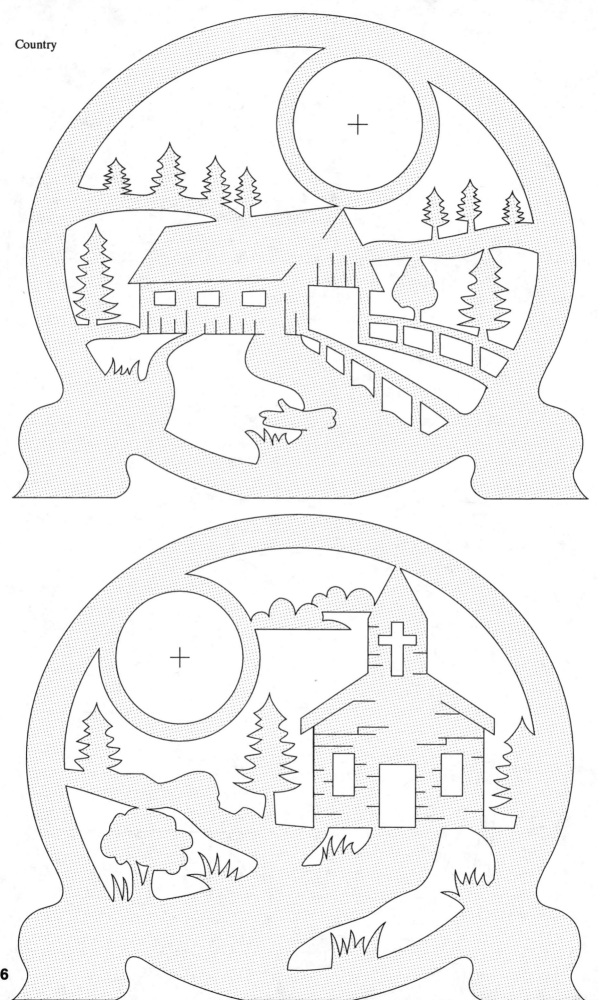

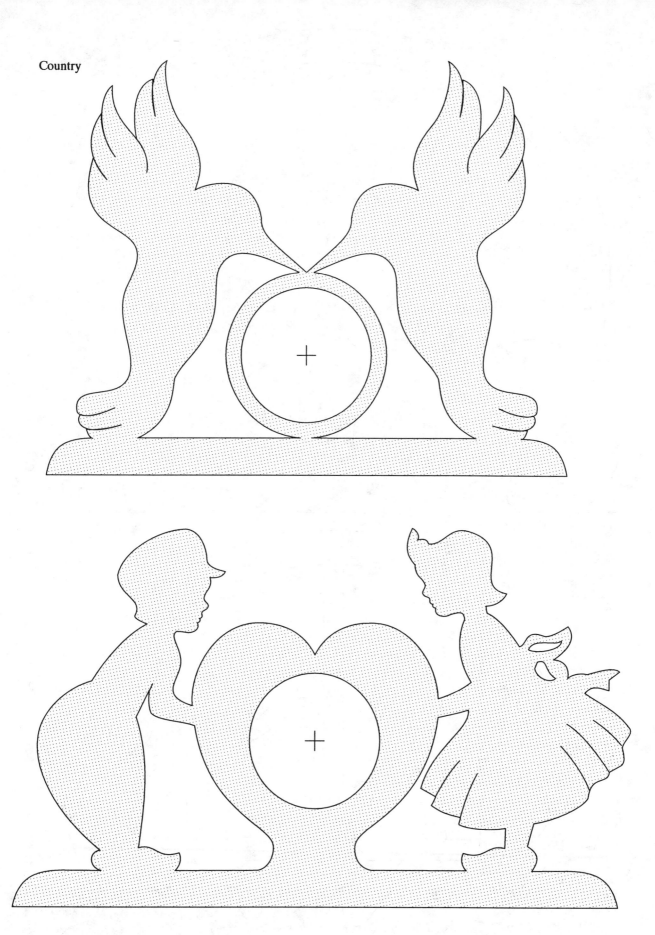

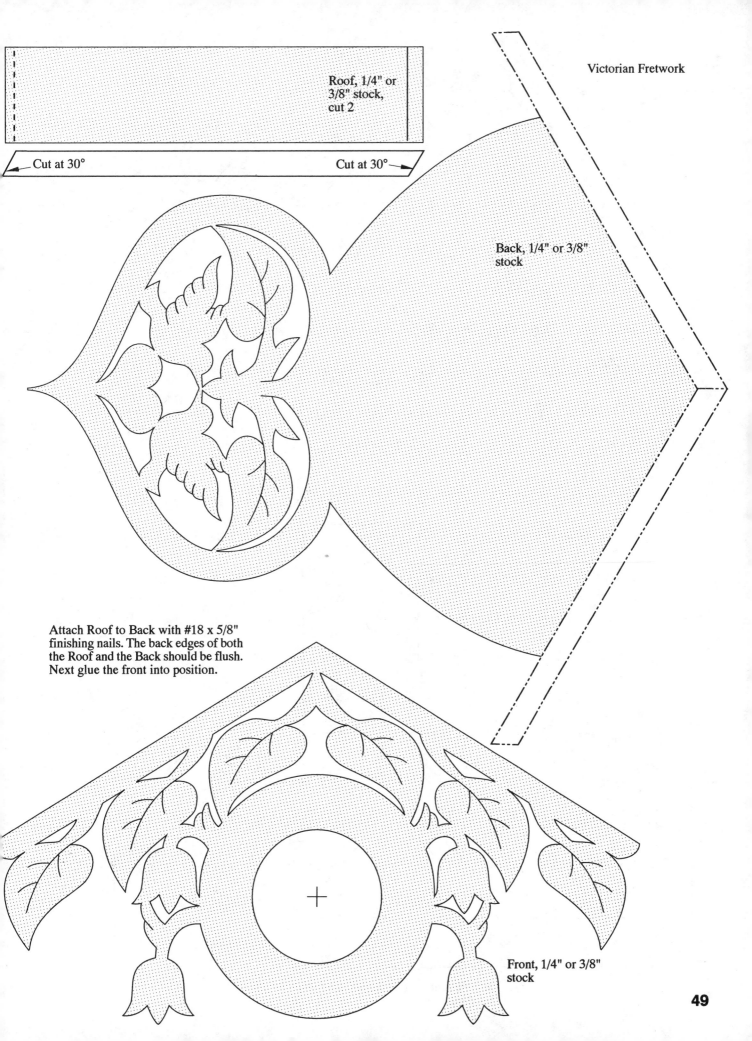

Roof, 1/4" or 3/8" stock, cut 2

Cut at 30°   Cut at 30°

Victorian Fretwork

Back, 1/4" or 3/8" stock

Attach Roof to Back with #18 x 5/8" finishing nails. The back edges of both the Roof and the Back should be flush. Next glue the front into position.

Front, 1/4" or 3/8" stock

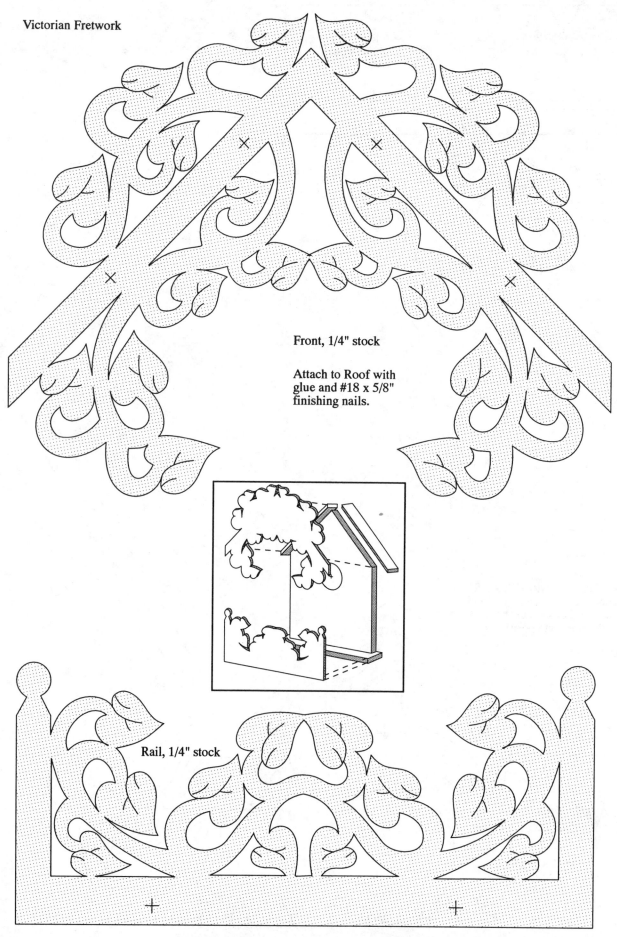

Front, 1/4" stock

Attach to Roof with glue and #18 x 5/8" finishing nails.

Rail, 1/4" stock

Attach to front of Base with glue and #18 x 5/8" finishing nails.

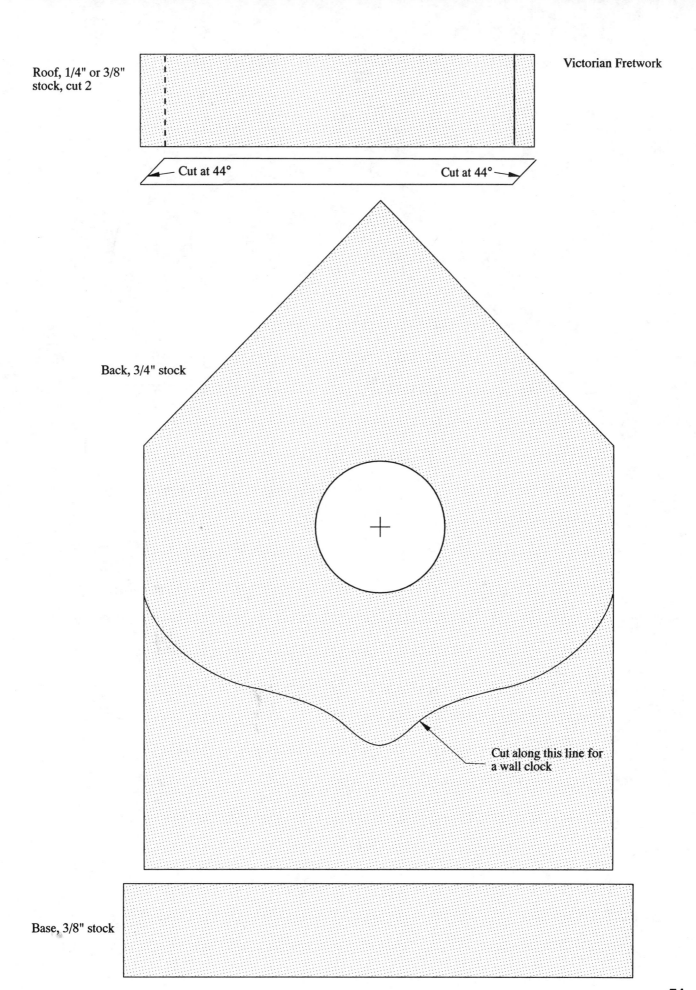

Roof, 1/4" or 3/8"
stock, cut 2

Victorian Fretwork

Cut at 44°   Cut at 44°

Back, 3/4" stock

Cut along this line for
a wall clock

Base, 3/8" stock

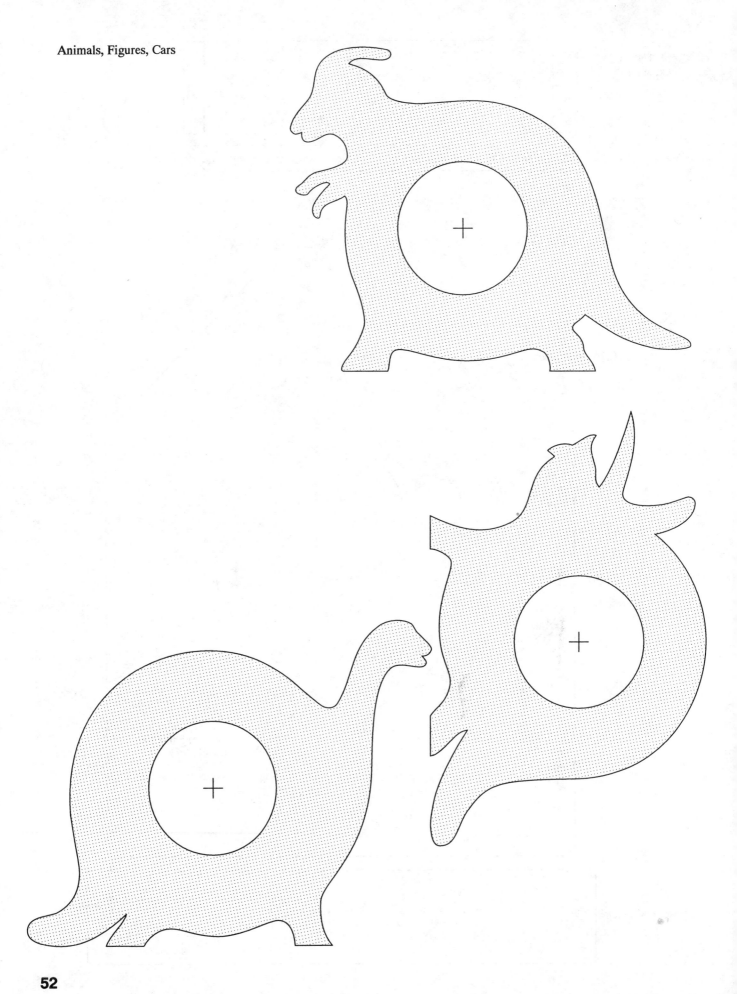

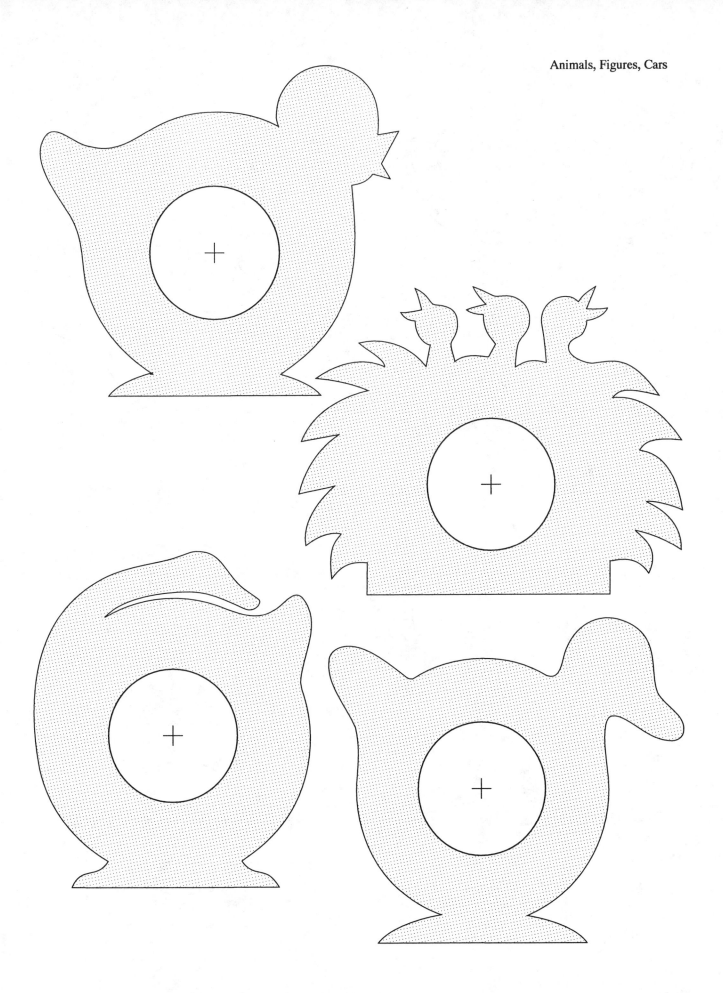

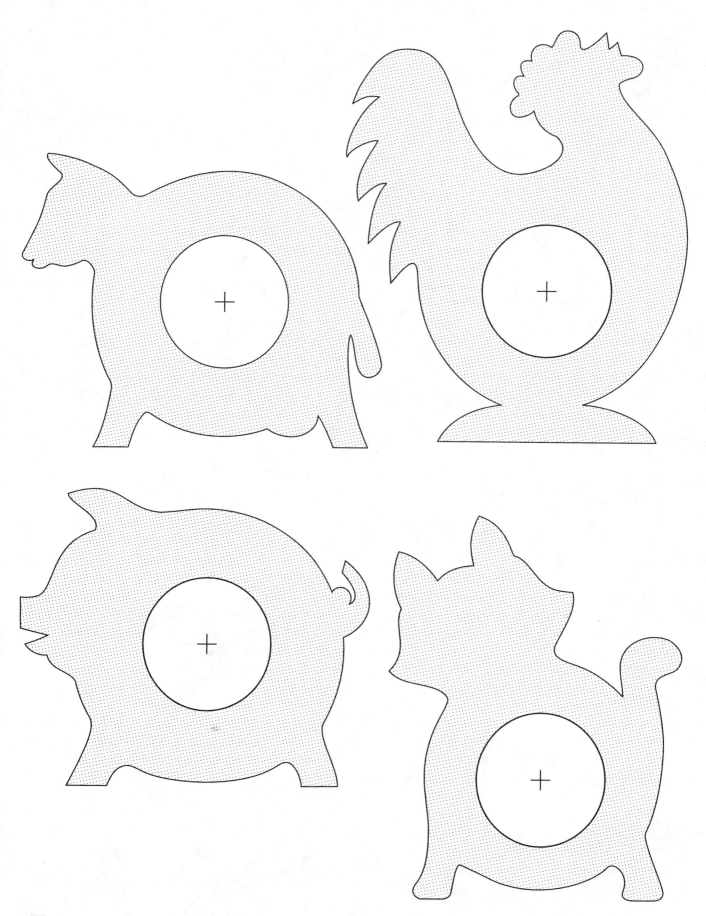

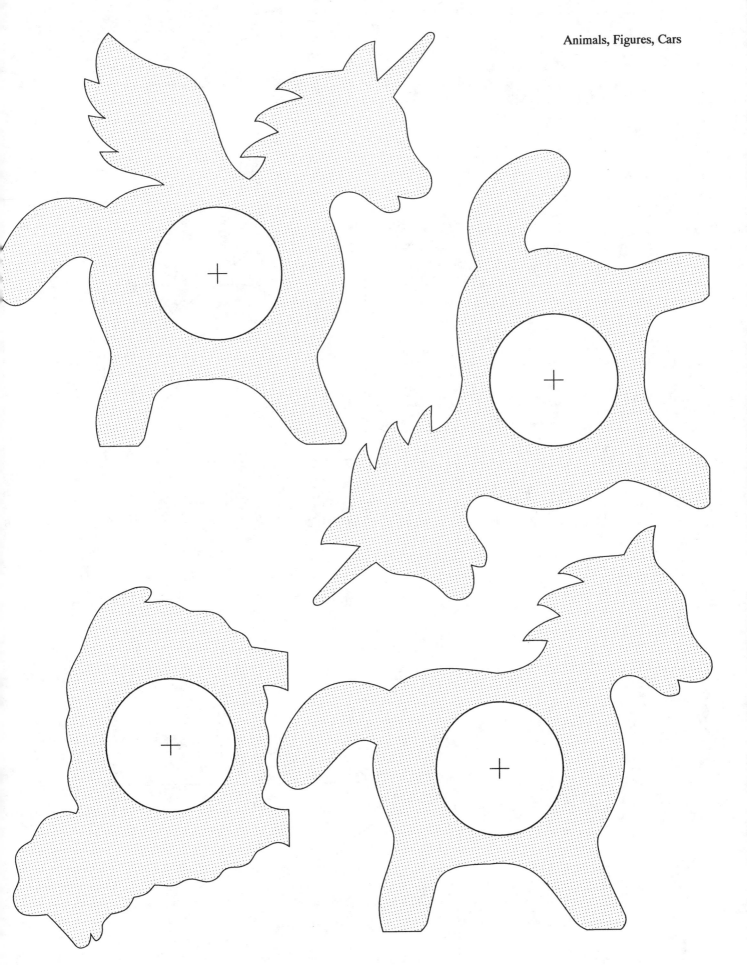

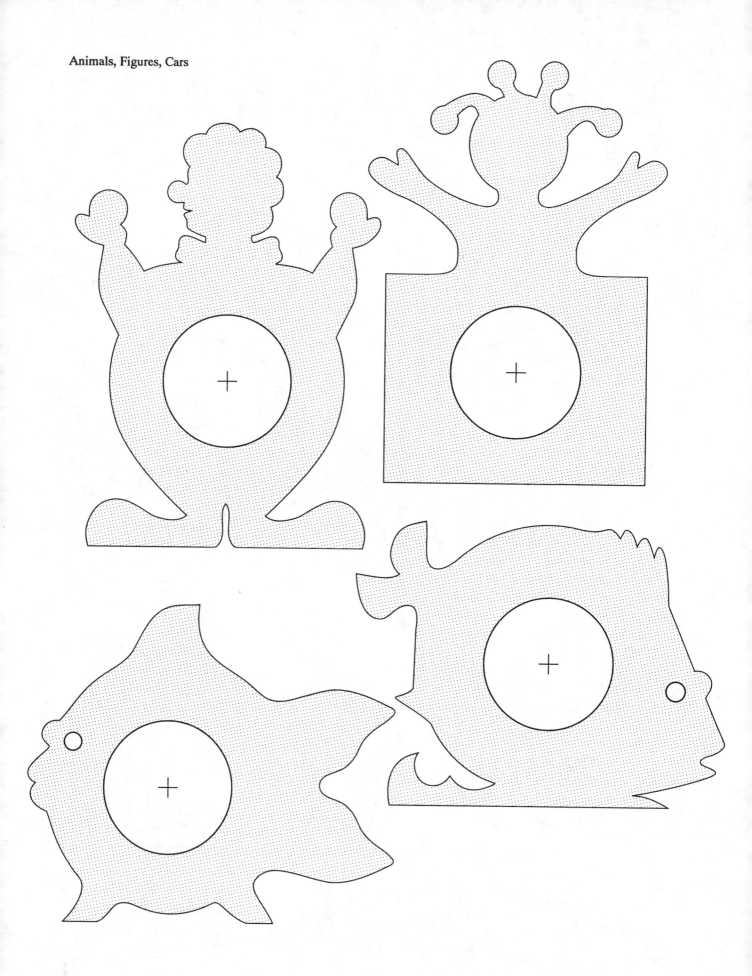

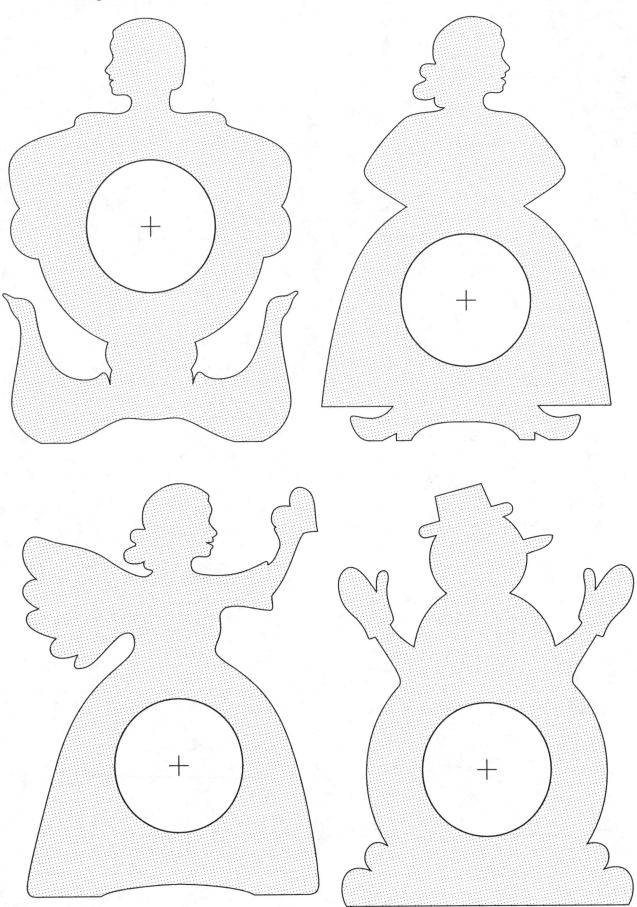

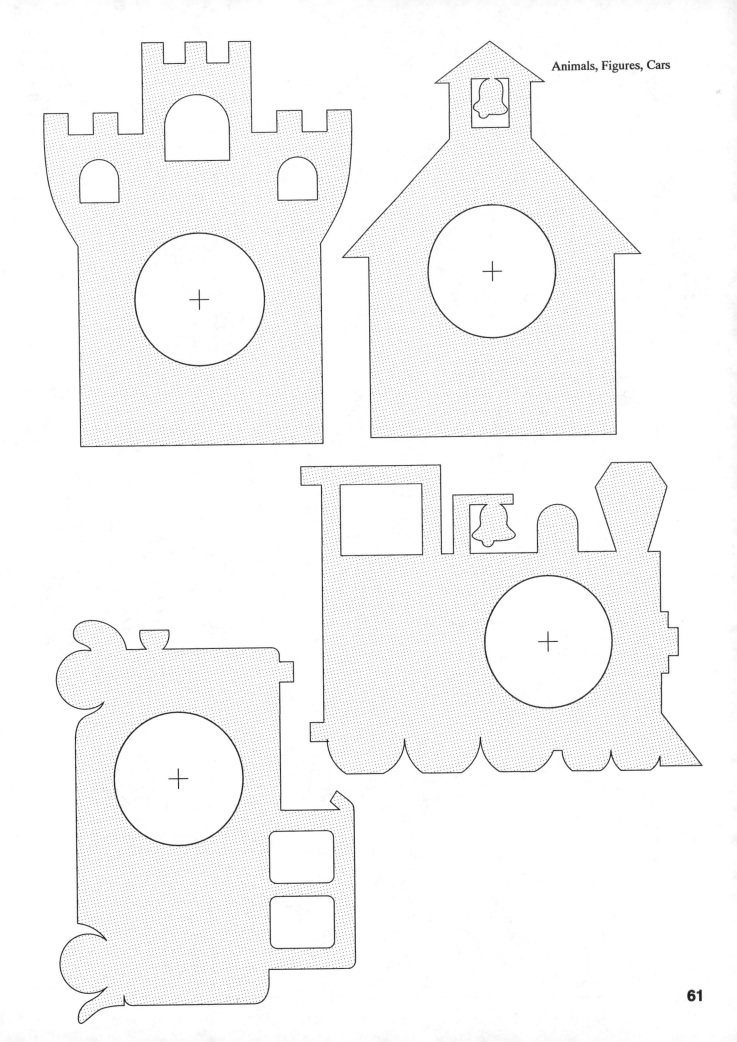

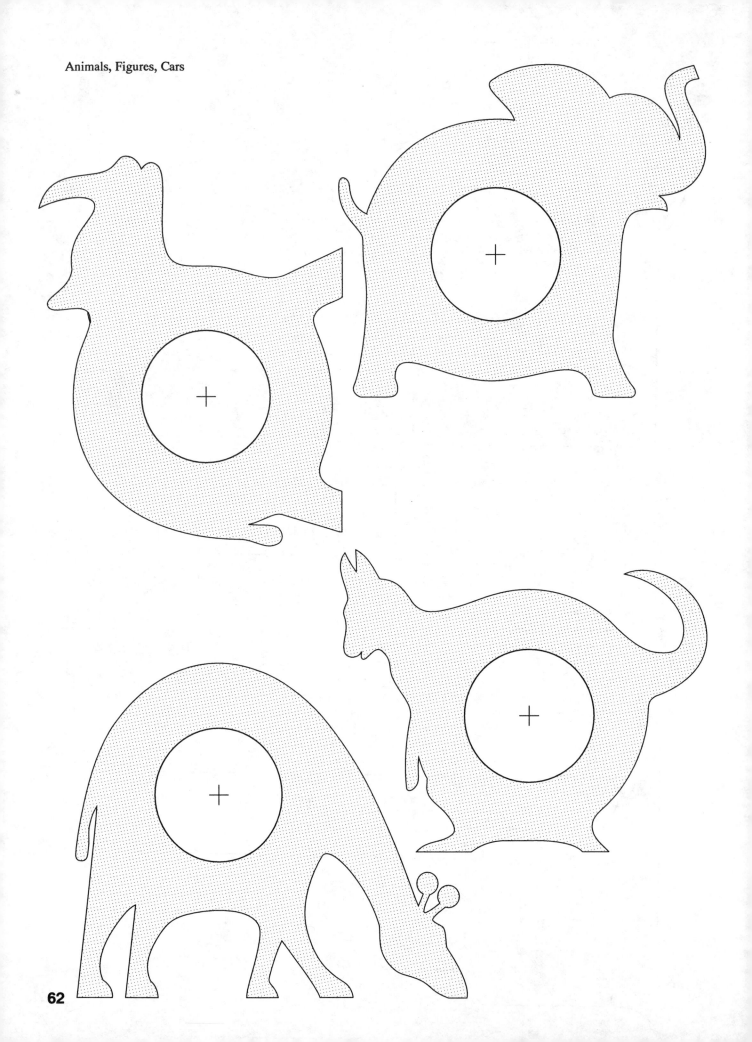

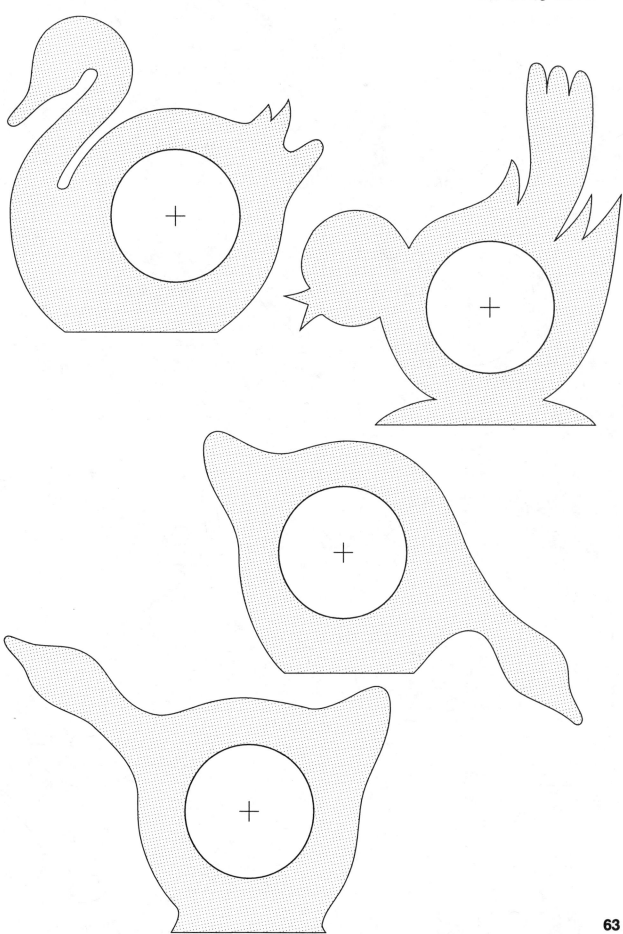

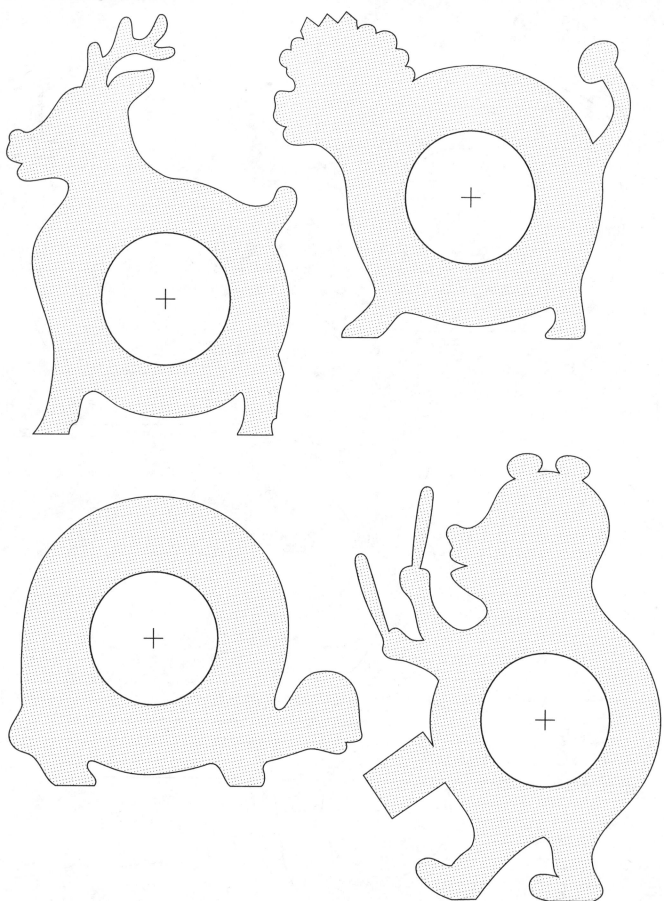

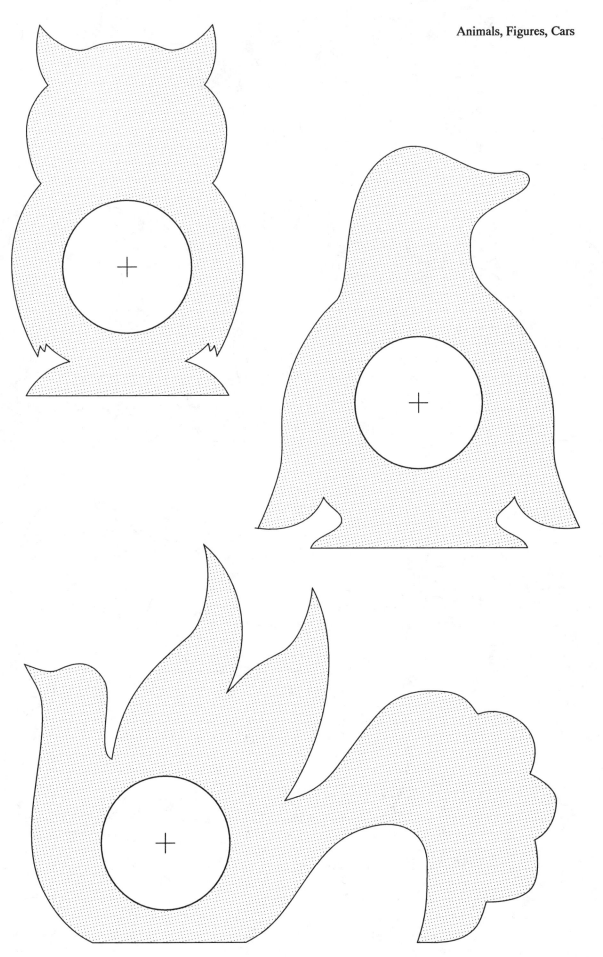

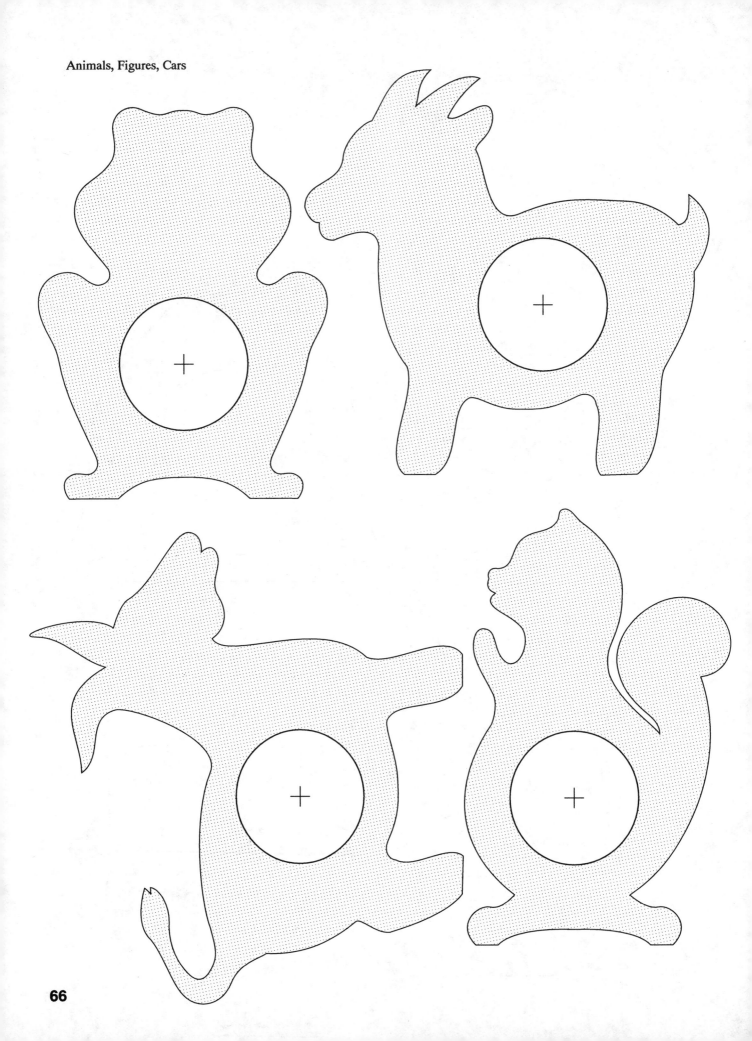

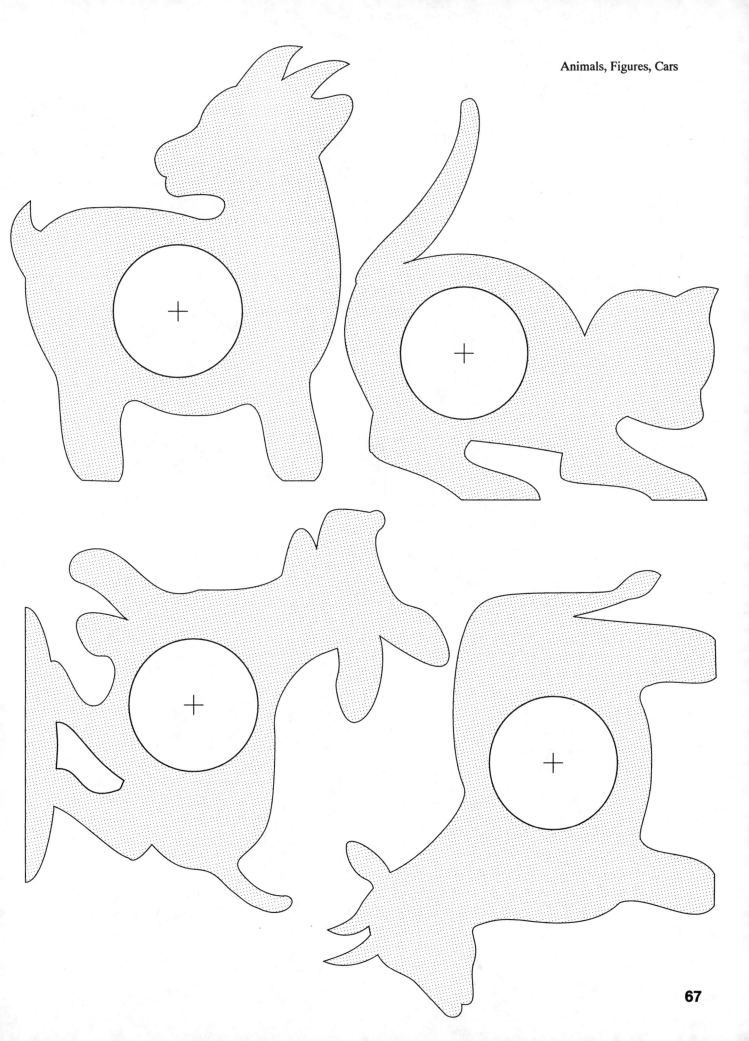

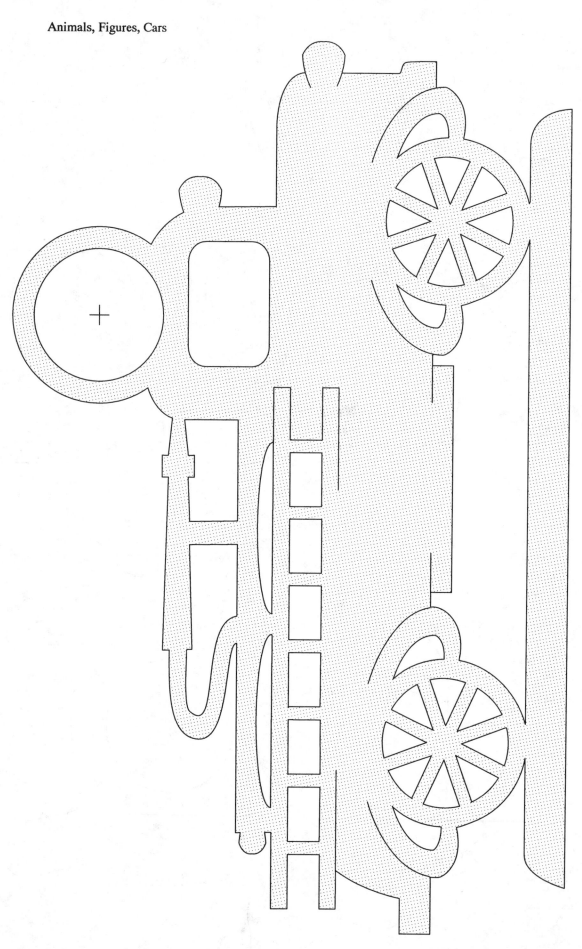

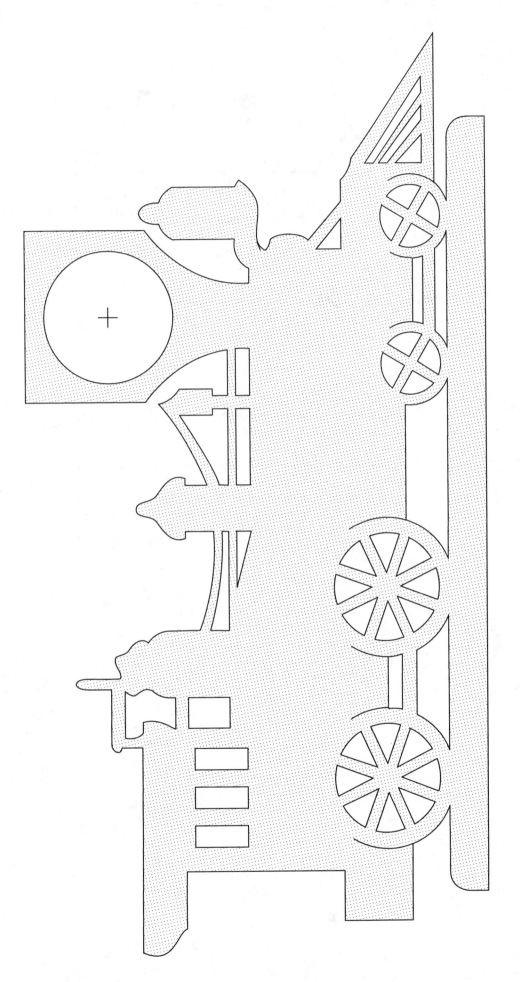

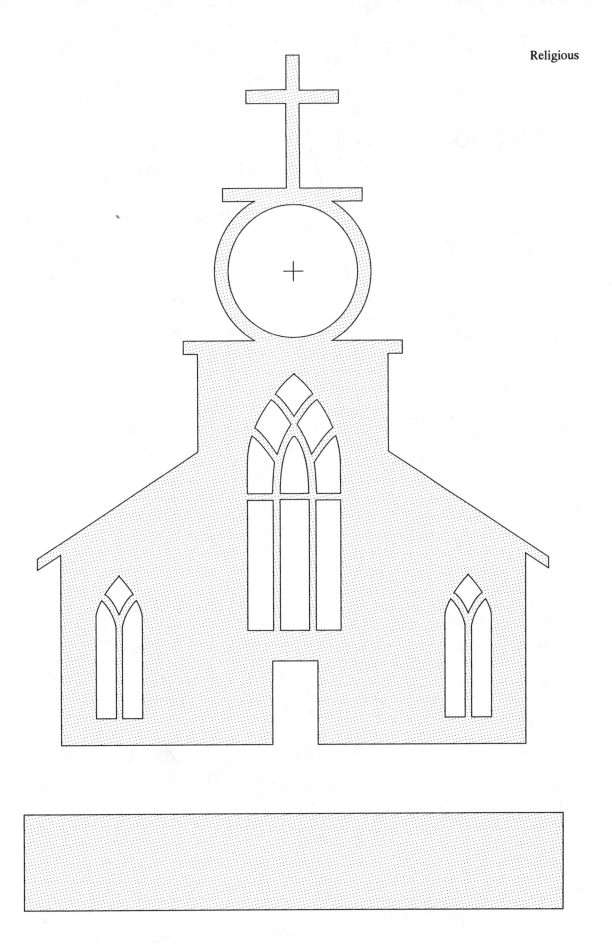

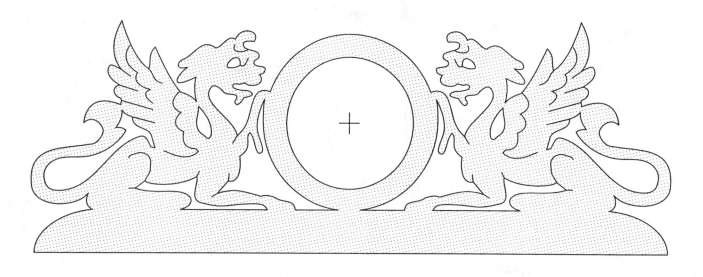

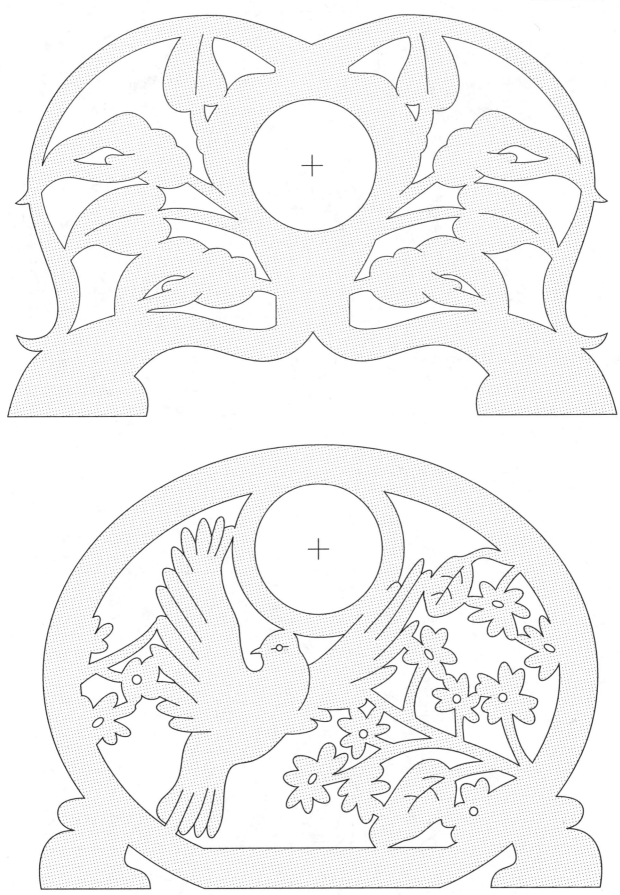

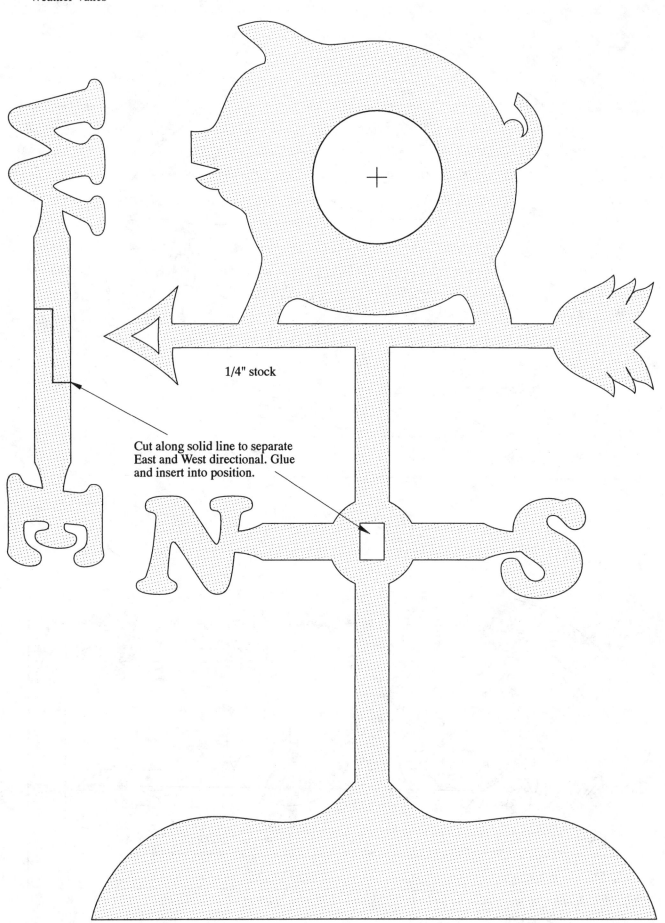

1/4" stock

Cut along solid line to separate
East and West directional. Glue
and insert into position.

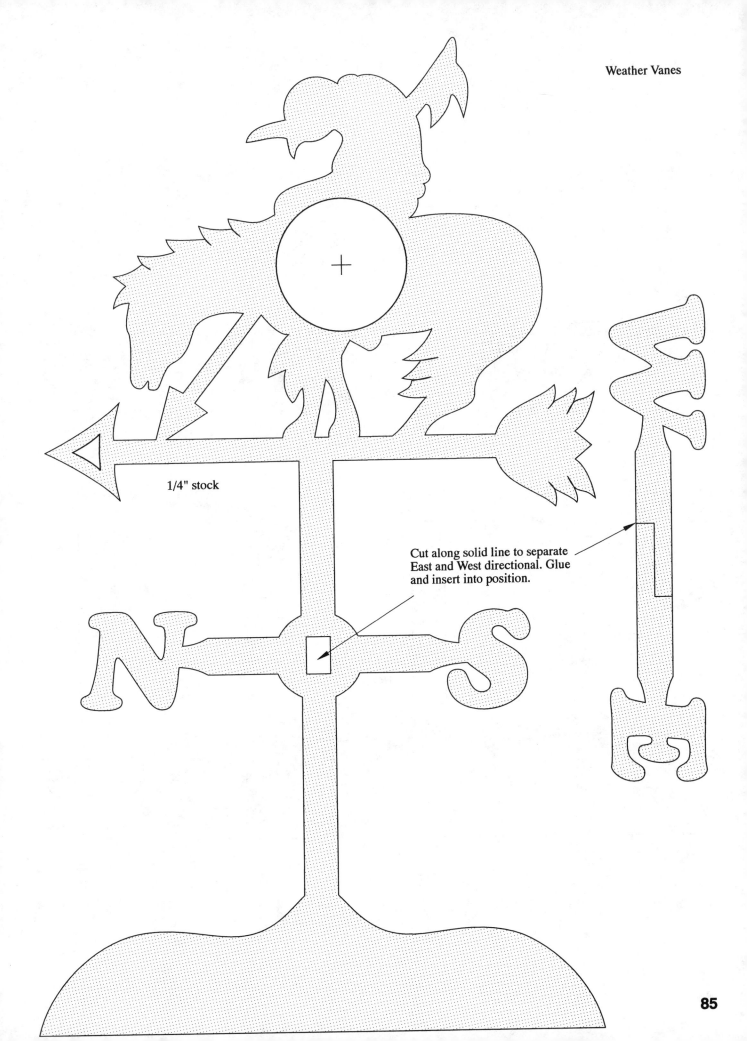

1/4" stock

Cut along solid line to separate
East and West directional. Glue
and insert into position.

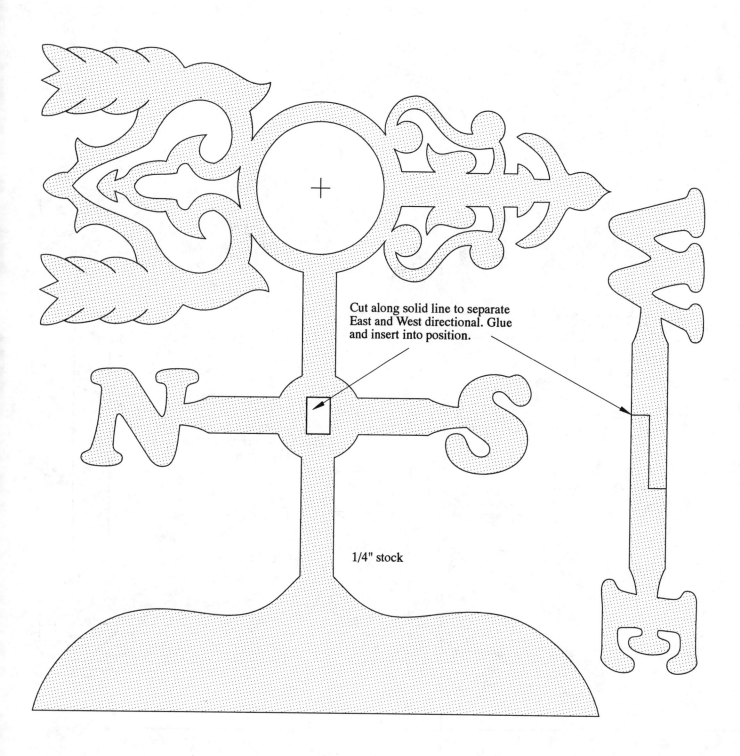

Cut along solid line to separate
East and West directional. Glue
and insert into position.

1/4" stock

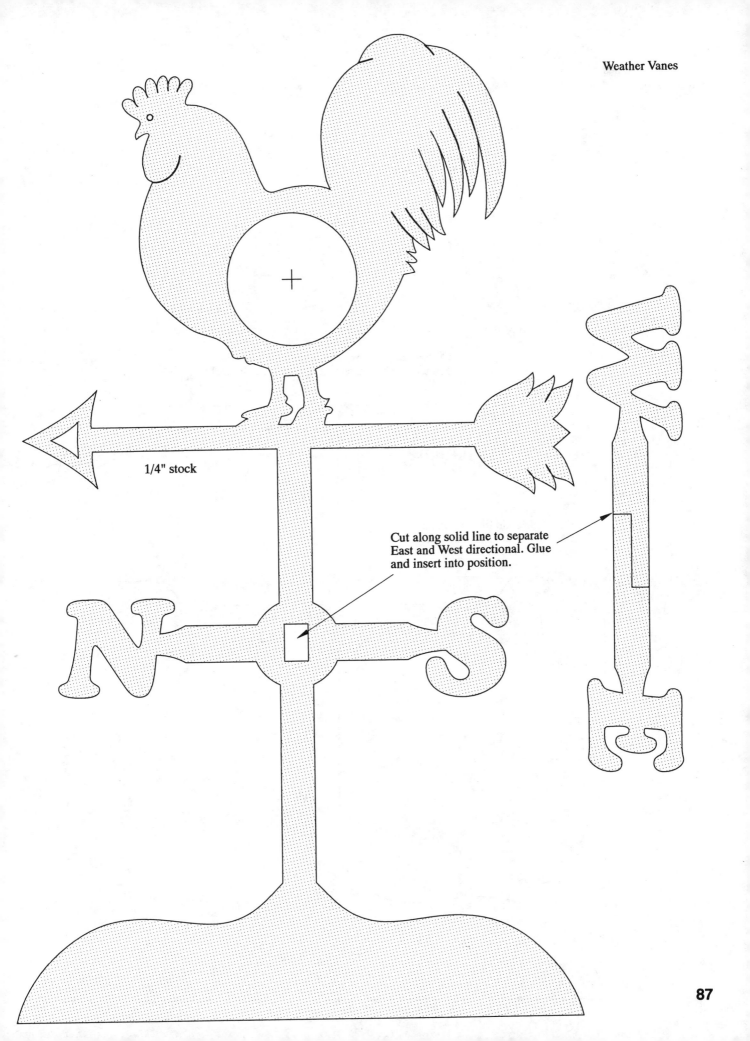

1/4" stock

Cut along solid line to separate East and West directional. Glue and insert into position.

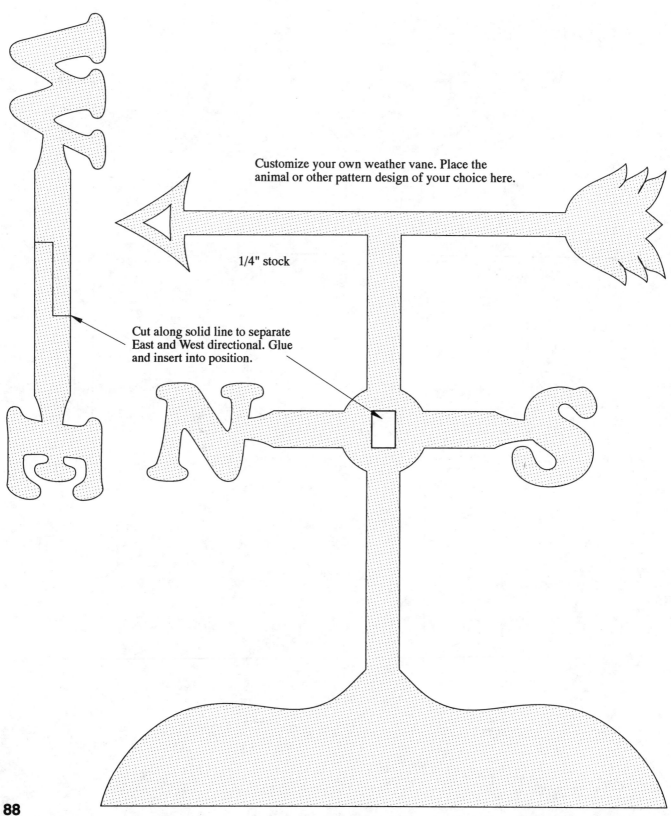

Customize your own weather vane. Place the
animal or other pattern design of your choice here.

1/4" stock

Cut along solid line to separate
East and West directional. Glue
and insert into position.

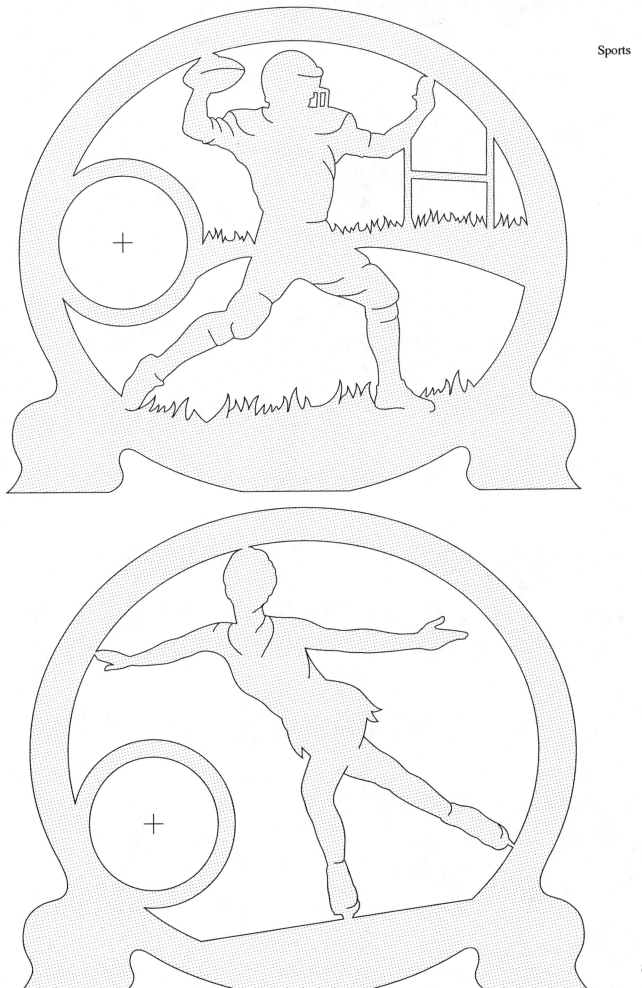

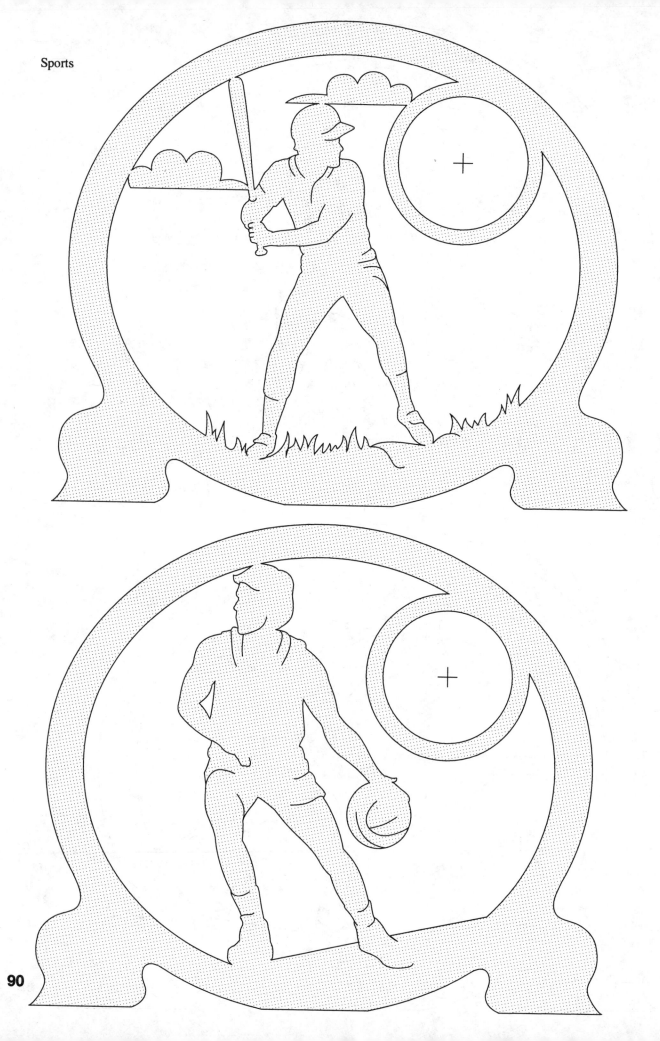

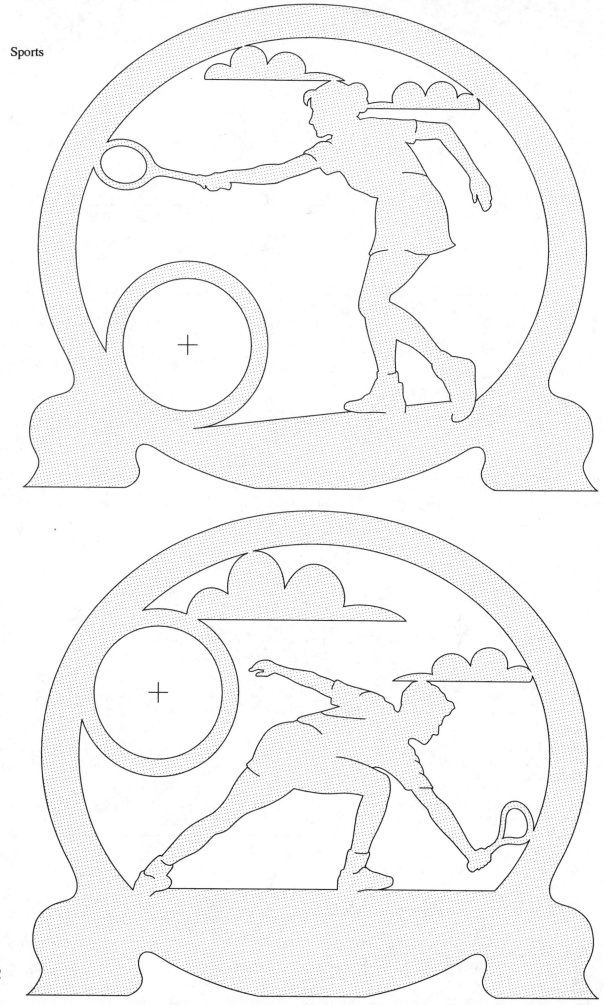

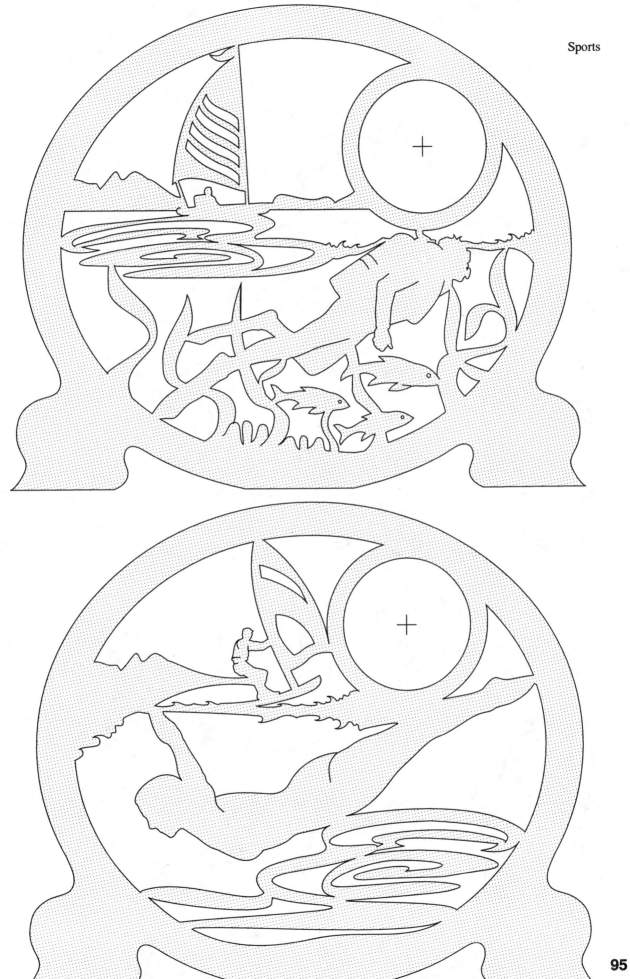

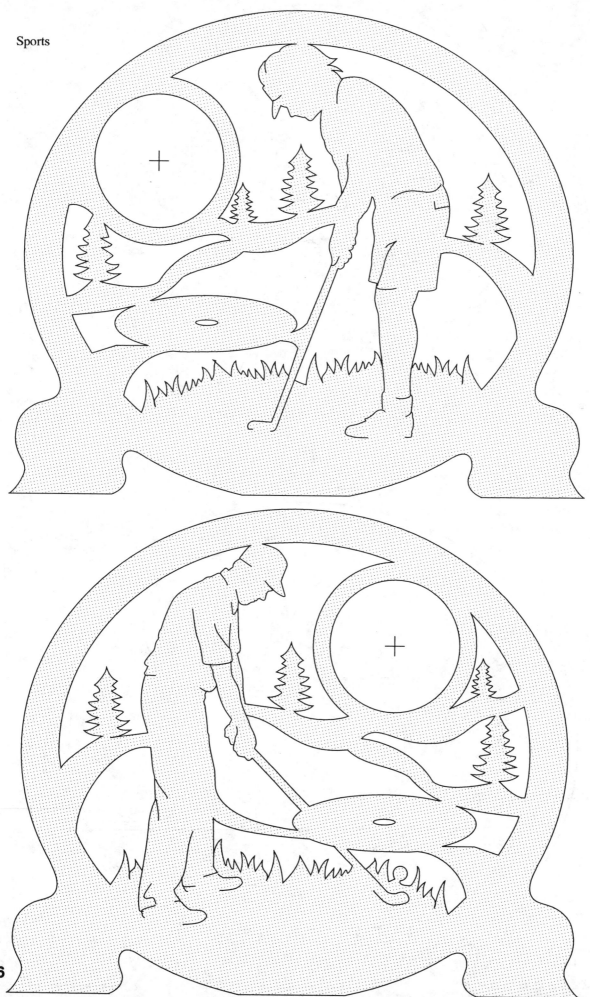

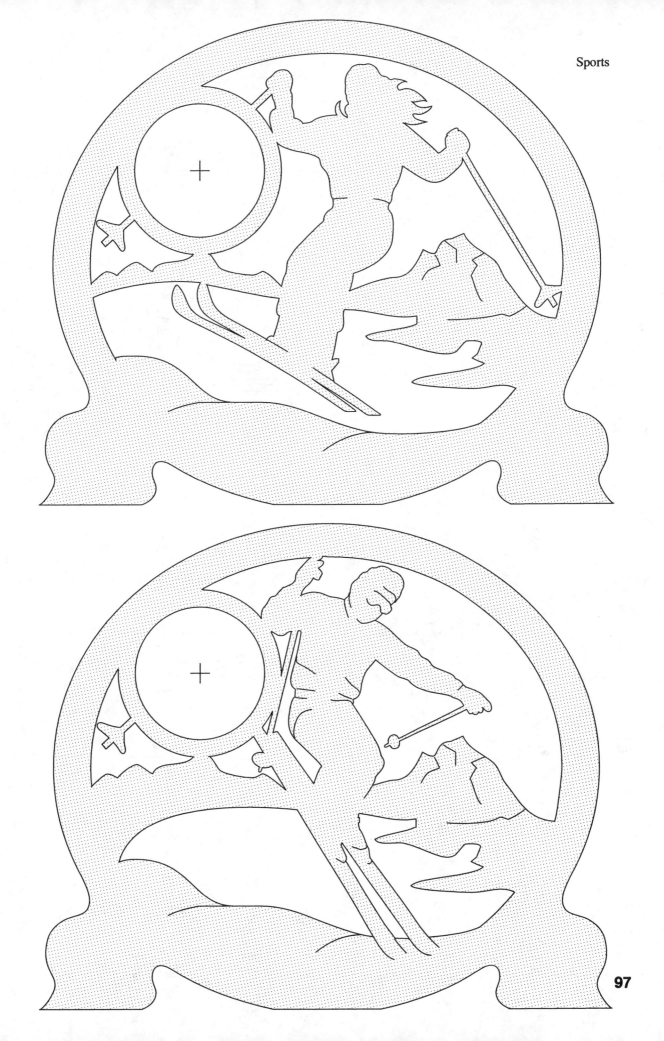

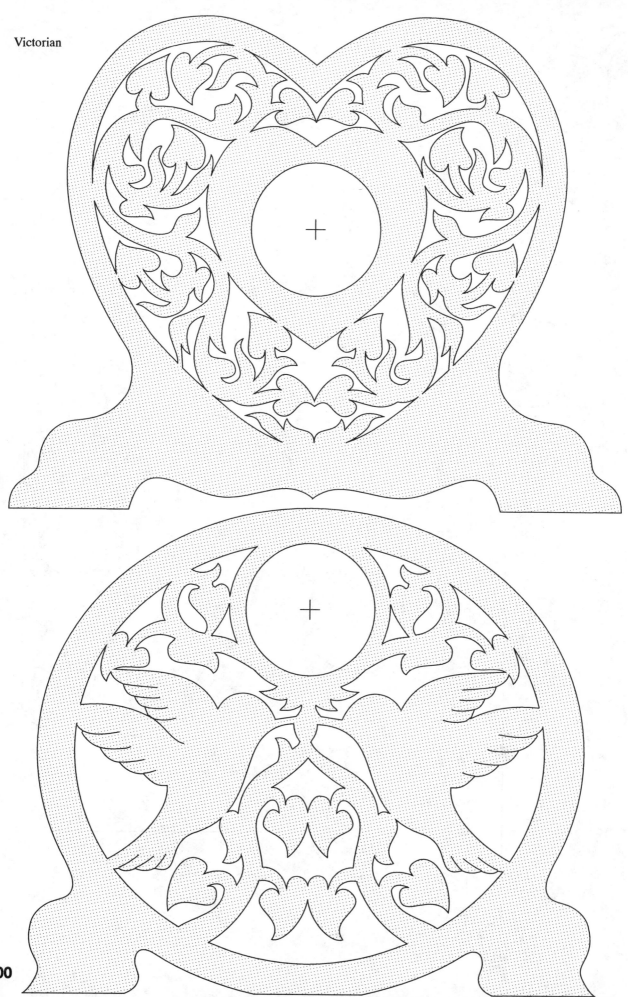

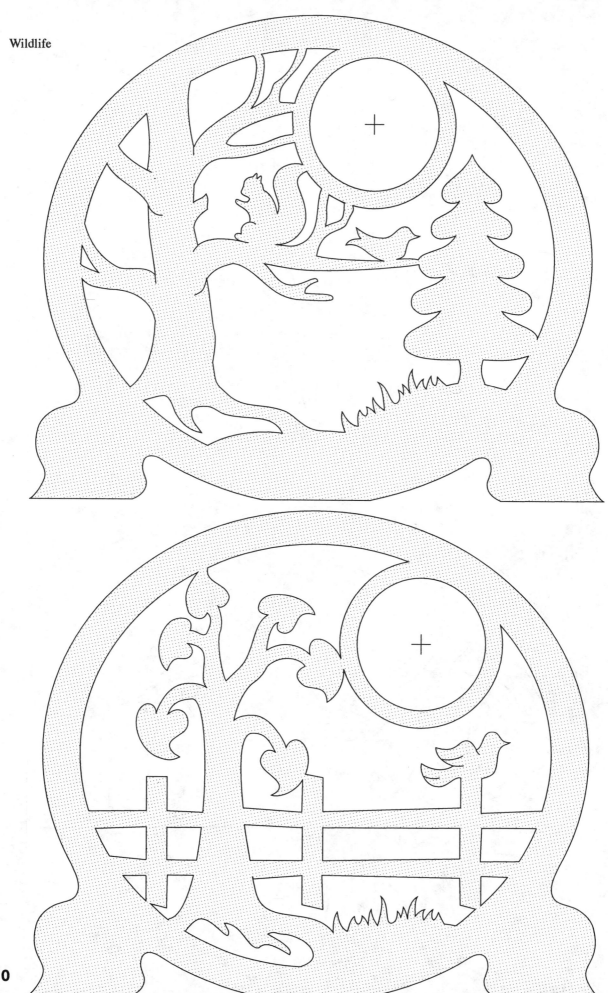

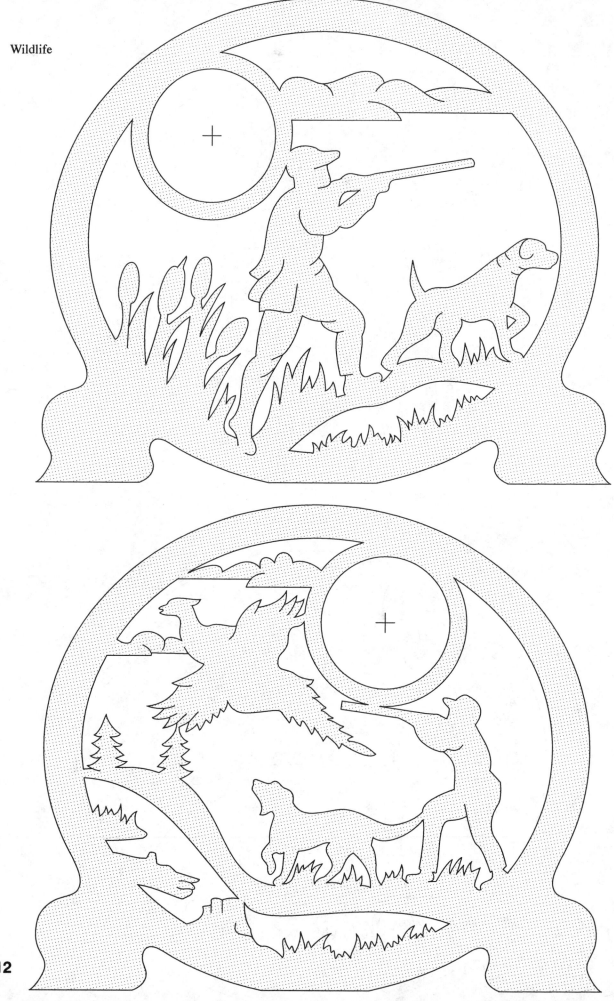

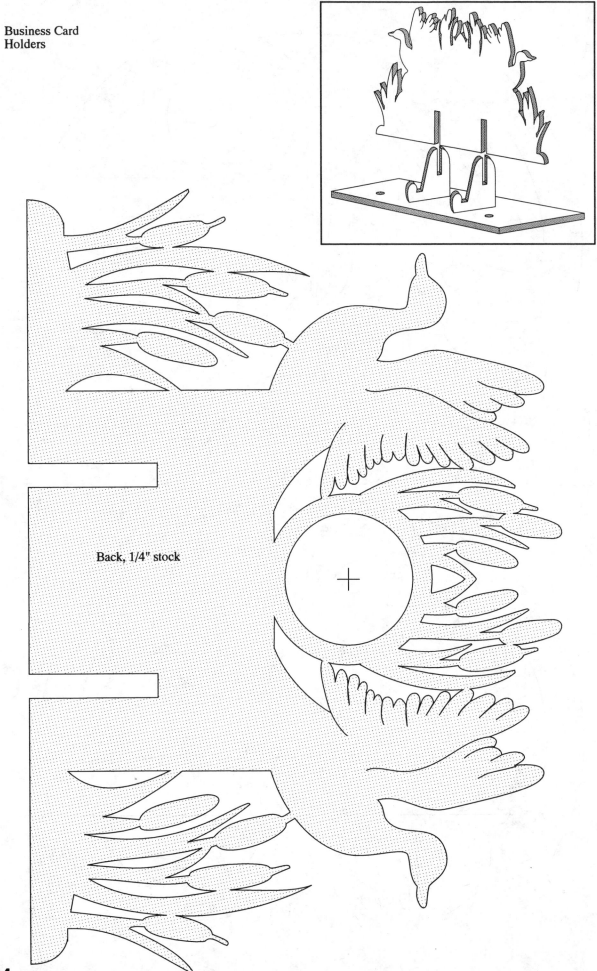

Back, 1/4" stock

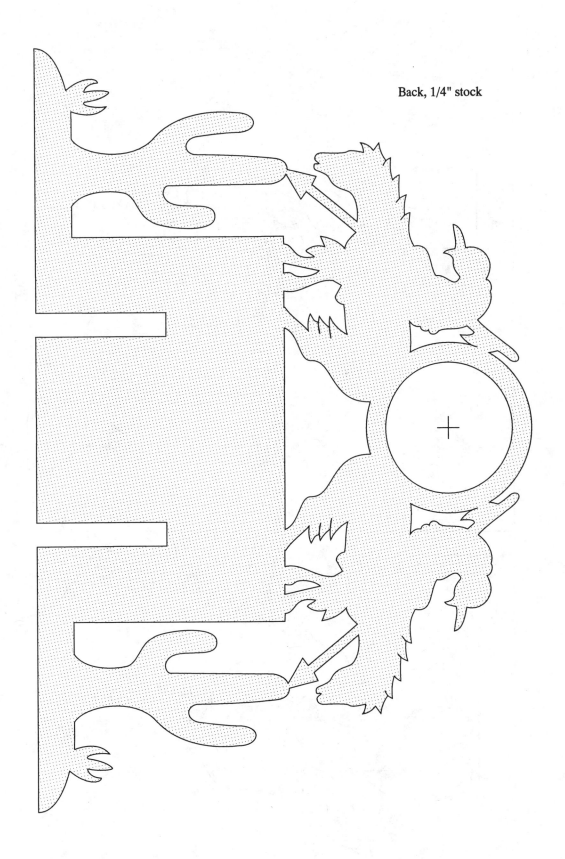

Back, 1/4" stock

Back, 1/4" stock

Back, 1/4" stock

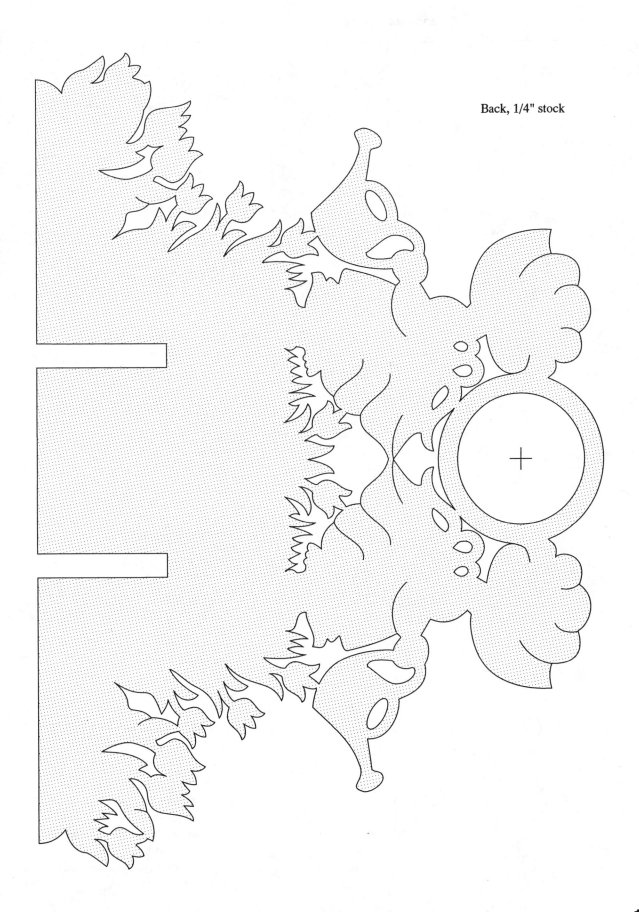

Base, 1/2" stock

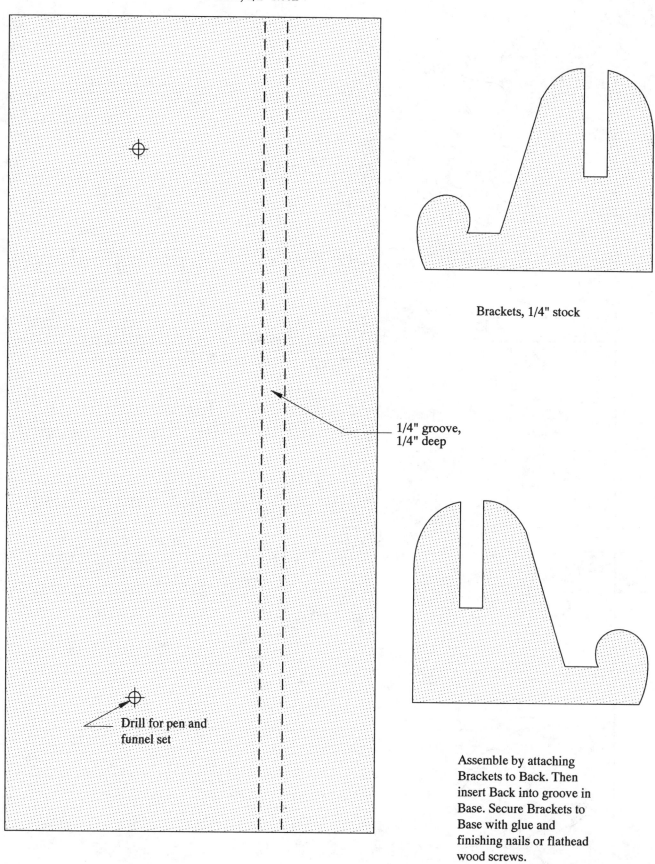

Brackets, 1/4" stock

1/4" groove,
1/4" deep

Drill for pen and
funnel set

Assemble by attaching
Brackets to Back. Then
insert Back into groove in
Base. Secure Brackets to
Base with glue and
finishing nails or flathead
wood screws.

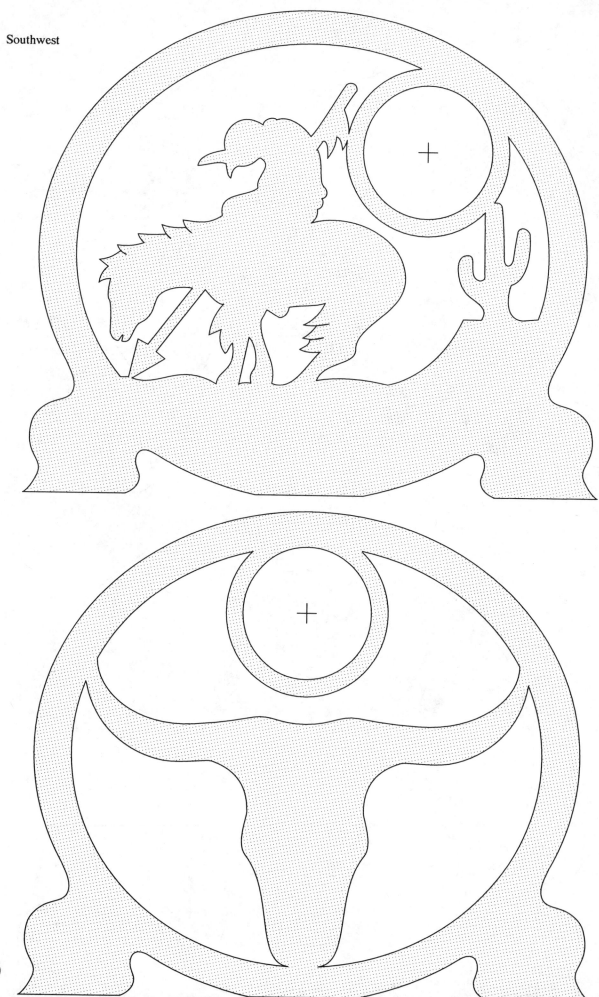

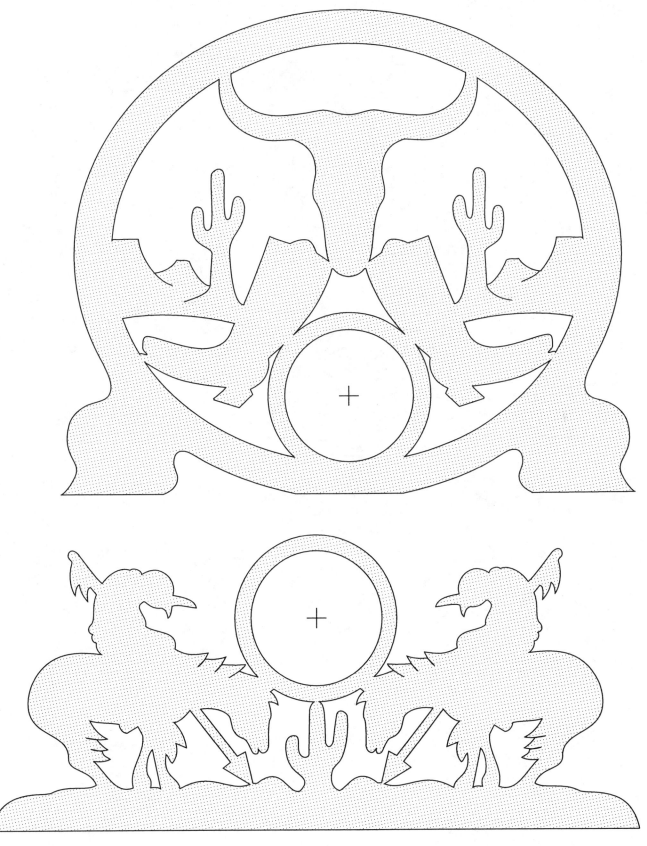

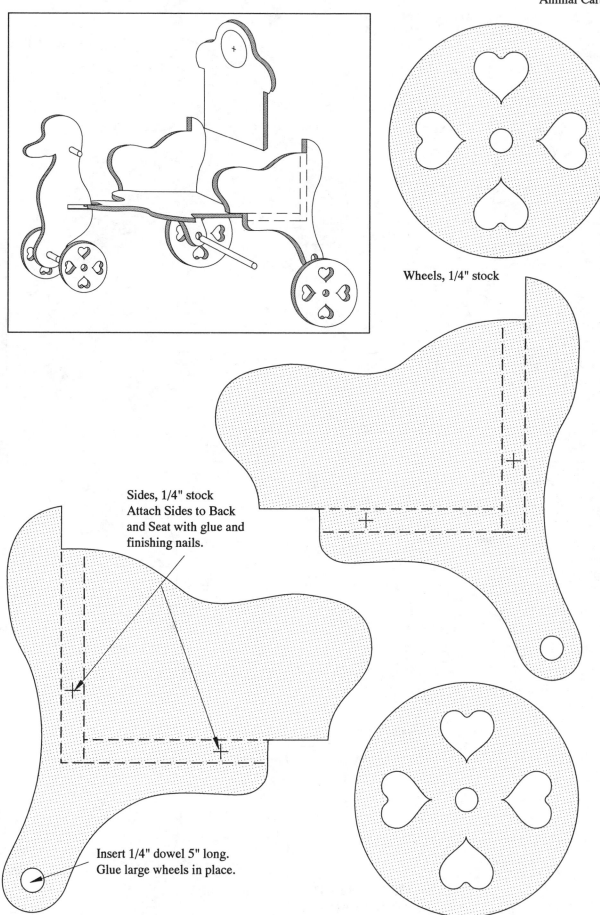

Wheels, 1/4" stock

Sides, 1/4" stock
Attach Sides to Back
and Seat with glue and
finishing nails.

Insert 1/4" dowel 5" long.
Glue large wheels in place.

Animal Carts

Wheels, 1/4" stock

Back, 1/4" stock

Head, 1/4" stock

Insert 1/4"
dowel 2" long.

Seat, 1/4" stock

Insert 1/4" dowel 1 1/4"
long. Glue small wheels
in place.

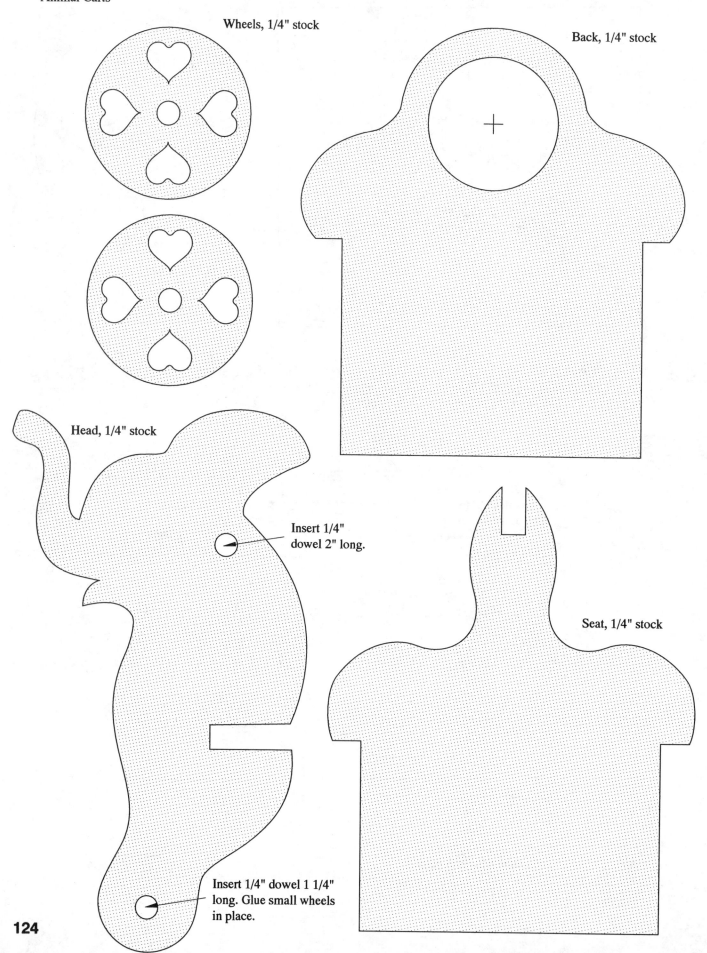

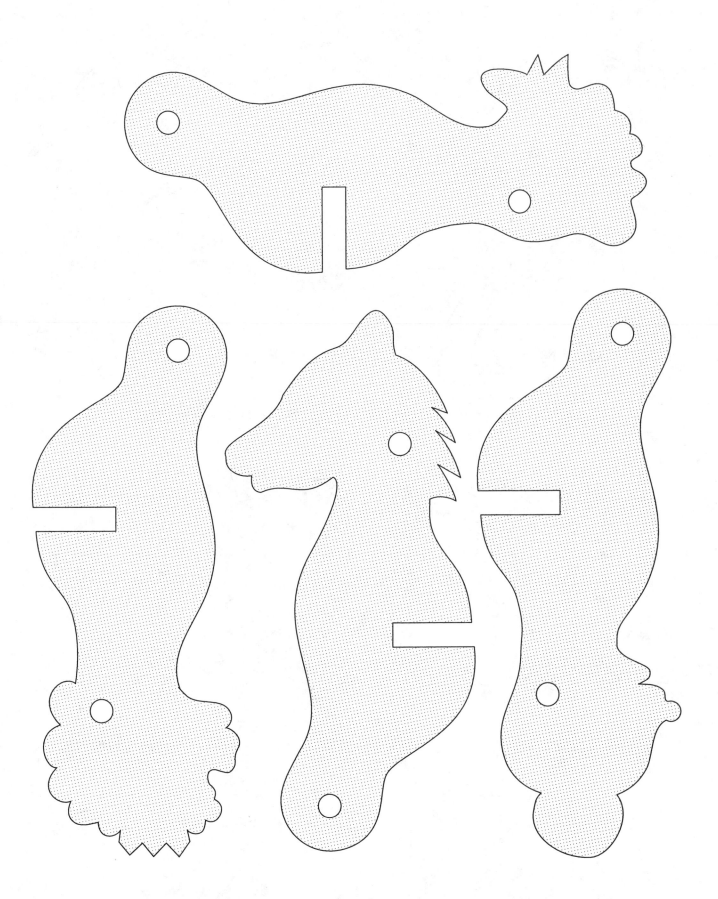

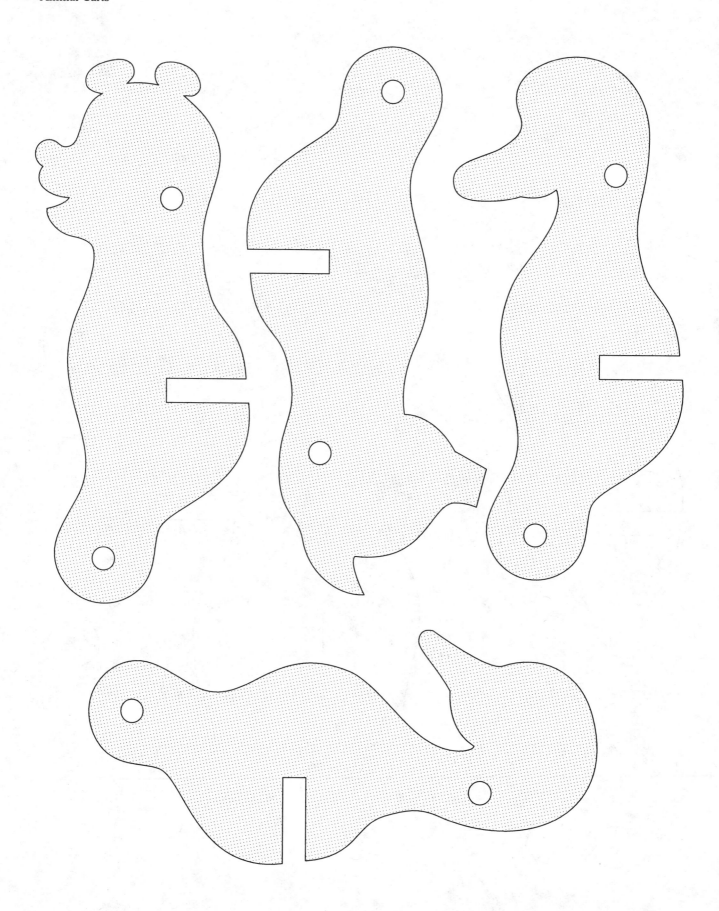

FREE CATALOG OFFER

Are you interested in more unique designs?

Yes, please add my name to your mailing list for a catalog of more unique ideas.

NAME _____

ADDRESS _____

CITY _____ STATE _____ ZIP _____

THE BERRY BASKET • PO BOX 925-BK4 • CENTRALIA, WA 98531 • 1-800-206-9009

Do you have friends who are interested in a catalog of unique ideas?

Yes, please add my friends to your mailing list and send them a catalog of unique ideas.

NAME _____

ADDRESS _____

CITY _____ STATE _____ ZIP _____

THE BERRY BASKET • PO BOX 925-BK4 • CENTRALIA, WA 98531 • 1-800-206-9009

WOODWORKING SURVEY

We always appreciate when people take time to write and let us know what they like and what they'd like to see more of. We know more of you would like to do the same, but find it hard to find the time. So here's an opportunity to do just that! It's easy - just grab a pen and mark a box! We'll personally look at every survey returned and use your responses to help design future patterns and projects.

1. My skill level is:
☐ Beginner ☐ Intermediate ☐ Advanced

2. I like to complete projects that are:
☐ Simple ☐ Intermediate ☐ Intricate

3. I prefer projects that require:
☐ Thin material ☐ Thick material (3/4" or more)
☐ Both

4. I feel the amount of instructions pertaining to the patterns in this book are:
☐ Clear and sufficient
☐ Unclear and incomplete

5. My favorite pattern themes are: (mark all that apply)

☐ Wildlife ☐ Religious
☐ Country ☐ Floral
☐ Victorian ☐ Sports
☐ Southwest ☐ Holiday/Celebration
☐ Children's ☐ Other _____

6. I would like more of the folowing projects:
(mark all that apply)

☐ Clocks ☐ Baskets
☐ Shelves ☐ Mirrors/Picture Frames
☐ Doll Furniture ☐ Plaques
☐ Birdhouses ☐ Other _____

Staple or tape to prepare for mail

Fold here second

- -

The Berry Basket
PO Box 925-BK4
Centralia, WA 98531

Stamp

The Berry Basket
PO Box 925-BK4
Centralia, WA 98531

- -

Fold here first

Staple or tape to prepare for mail